Mathematical Models of Crop Growth and Yield

Soil Biochemistry, Volume 1, edited by A. D. McLaren and G. H. Peterson
Soil Biochemistry, Volume 2, edited by A. D. McLaren and J. Skujiņš
Soil Biochemistry, Volume 3, edited by E. A. Paul and A. D. McLaren
Soil Biochemistry, Volume 4, edited by E. A. Paul and A. D. McLaren
Soil Biochemistry, Volume 5, edited by E. A. Paul and J. N. Ladd
Soil Biochemistry, Volume 6, edited by Jean-Marc Bollag and G. Stotzky
Soil Biochemistry, Volume 7, edited by G. Stotzky and Jean-Marc Bollag
Soil Biochemistry, Volume 8, edited by Jean-Marc Bollag and G. Stotzky
Soil Biochemistry, Volume 9, edited by G. Stotzky and Jean-Marc Bollag
Soil Biochemistry, Volume 10, edited by Jean-Marc Bollag and G. Stotzky

Organic Chemicals in the Soil Environment, Volumes 1 and 2, edited by C. A. I. Goring and J. W. Hamaker
Humic Substances in the Environment, M. Schnitzer and S. U. Khan
Microbial Life in the Soil: An Introduction, T. Hattori
Principles of Soil Chemistry, Kim H. Tan
Soil Analysis: Instrumental Techniques and Related Procedures, edited by Keith A. Smith
Soil Reclamation Processes: Microbiological Analyses and Applications, edited by Robert L. Tate III and Donald A. Klein
Symbiotic Nitrogen Fixation Technology, edited by Gerald H. Elkan
Soil–Water Interactions: Mechanisms and Applications, Shingo Iwata and Toshio Tabuchi with Benno P. Warkentin

The Rhizosphere: Biochemistry and Organic Substances at the Soil–Plant Interface, Roberto Pinton, Zeno Varanini, and Paolo Nannipieri

Woody Plants and Woody Plant Management: Ecology, Safety, and Environmental Impact, Rodney W. Bovey

Metals in the Environment: Analysis by Biodiversity, M. N. V. Prasad

Plant Pathogen Detection and Disease Diagnosis: Second Edition, Revised and Expanded, P. Narayanasamy

Handbook of Plant and Crop Physiology: Second Edition, Revised and Expanded, edited by Mohammad Pessarakli

Environmental Chemistry of Arsenic, edited by William T. Frankenberger, Jr.

Enzymes in the Environment: Activity, Ecology, and Applications, edited by Richard G. Burns and Richard P. Dick

Plant Roots: The Hidden Half, Third Edition, Revised and Expanded, edited by Yoav Waisel, Amram Eshel, and Uzi Kafkafi

Handbook of Plant Growth: pH as the Master Variable, edited by Zdenko Rengel

Biological Control of Crop Diseases, edited by Samuel S. Gnanamanickam

Pesticides in Agriculture and the Environment, edited by Willis B. Wheeler

Mathematical Models of Crop Growth and Yield, Allen R. Overman and Richard V. Scholtz III

Plant Biotechnology and Transgenic Plants, edited by Kirsi-Marja Oksman-Caldentey and Wolfgang H. Barz

Additional Volumes in Preparation

Handbook of Postharvest Technology, edited by A. Chakraverty, Arun S. Mujumdar, G. S. V. Raghavan, and H. S. Ramaswamy

Handbook of Soil Acidity, edited by Zdenko Rengel

MATHEMATICAL MODELS OF CROP GROWTH AND YIELD

ALLEN R. OVERMAN
RICHARD V. SCHOLTZ III

University of Florida
Gainesville, Florida, U.S.A.

CRC Press
Taylor & Francis Group
Boca Raton London New York

CRC Press is an imprint of the
Taylor & Francis Group, an **informa** business

Cover photograph: Michael D. Minichiello.

CRC Press
Taylor & Francis Group
6000 Broken Sound Parkway NW, Suite 300
Boca Raton, FL 33487-2742

First issued in paperback 2019

© 2002 by Taylor & Francis Group, LLC
CRC Press is an imprint of Taylor & Francis Group, an Informa business

No claim to original U.S. Government works

ISBN-13: 978-0-8247-0825-2 (hbk)
ISBN-13: 978-0-367-39589-6 (pbk)

Visit the Taylor & Francis Web site at
http://www.taylorandfrancis.com

and the CRC Press Web site at
http://www.crcpress.com

Preface

This book is intended to outline an approach to crop modeling that I have found to be both mathematically solid and feasible to use in practice. My strategy is to develop the technical details in a way that offers some insight into a logical progression from a simple idea toward more complex details. The reader should sense something of the uncertainty involved in the struggle to interpret data and compose models for describing empirical results. My search has been driven by a combination of practical need and intellectual curiosity.

Science can be viewed as the search for: (1) patterns, (2) relationships, (3) connections, (4) consistency, and (5) beauty. Look for patterns (sometimes trends) in plots of data. Then search for mathematical relationships that agree with the patterns. Seek to identify connections between or among various factors in the analysis. Check for consistency from one data set to others (different investigators and different conditions). Finally, look for mathematical beauty (such as symmetry) in the models. The last is usually the most difficult to achieve and is often ignored.

The language of modern science is mathematics. This has become increasingly true in physics, chemistry, and biology. We can only marvel at how abstract ideas in mathematics have found application in science and engineering. Perhaps that is because many ideas were generalized from particular problems and observations. Attempts at quantitative descriptions of plant and animal systems inevitably lead to this same mathematical pathway. Unfortunately, many people are intimidated by the language and methods of applied mathematics. By the nature of the subject, this book necessarily involves many equations of different types. On a relative scale, the functions employed herein can be mastered with nominal effort and determination. The functions are all of the analytical type, in contrast to numerical techniques. Throughout I have followed the practice of starting with simpler models and then progressing towards greater levels of complexity and comphrehensiveness.

The reader is encouraged to scan the book to view data and model simulations before trying to master mathematical details, since the ultimate criterion for practical usefulness is agreement between the two. There is some repetition in the contents. Data are presented in both tabular and graphical form for several cases to help the reader obtain a more comprehensive grasp of the material. Graphs are sometimes shown in curvilinear form (Y vs. X) and then in linearized form (semilog or probability) to clarify trends and scatter in data. In addition, the same equation may be listed more than once in a chapter to enhance the flow of the exposition. Finally, in many cases the same symbol (e.g., A) is used with different meanings to avoid a plethora of symbols. I have tried to carefully define variables and parameters for each particular usage.

Childhood on a farm in eastern North Carolina gave me the opportunity to observe seasons and growth of plants and animals. This led to many questions about the process of farming. How did farmers know when to plant their crops? What was the relationship of crop growth to climate, water, and nutrients? It was obvious that all these factors influenced production. Was farming a craft, an art, or a science? In high school I found science courses more interesting than agricultural courses and was led to believe that some of the answers to my questions might lie within the framework of science.

As I read about engineering, there appeared many parallels between farming and engineering. As a result, in deciding on a major for college I chose agricultural engineering at North Carolina State College. Actually, my first exposure to this profession occurred at the age of 12 through a 4-H Club project in tractor maintenance. Eventually, I was to obtain B.S, M.S., and Ph.D. degrees in this field at North Carolina State University. The academic program was heavy in math, physics, and chemistry. Course work also included soil physics and soil chemistry but very little in plant science.

My interest in plant growth and production lay dormant until my postdoctorate work in agronomy at the University of Illinois. While there I audited a course on soil fertility taught by Professor Sigurd Melsted. Through his course I was introduced to the models of Mitsherlich and multivariate regression. It was very clear that these were empirical models without any physical basis. After joining the agricultural engineering faculty at the University of Florida my interest in crop models was stimulated by studies on crop response to waste application from agricultural and municipal sources. It was natural to invoke the Mitscherlich model, partly because of its mathematical simplicity and ease of calibration. Since it seemed to work reasonably well in relating crop yield to applied N, P, or K, I then performed an extensive review of numerous field studies to develop a broader perspective on the subject. It was during this study that I expanded the Mitscherlich equation to describe plant nutrient uptake as well as dry matter production. This seemed logical enough, and it appeared to describe the data reasonably well. However, the approach presented a paradox as soon as I tried to calculate plant nutrient concentration as the ratio of plant nutrient uptake to plant dry matter. In some cases the ratio either went negative or blew up at low applied nutrient levels! It was clear that there was a problem, but the question was how to either avoid the problem or fix it somehow. The subject was laid aside to incubate while attention was focused on other more pressing matters.

Progress came from a slightly different direction. I was challenged to either select or develop a dynamic model to relate plant growth (dry matter and plant nutrient accumulation) to time. Concerns with groundwater quality related to agriculture and waste management created pressure to incorporate practical models into the analysis and engineering design. By now I had an idea for a relatively simple mathematical approach. Analysis of some field data confirmed the usefulness of the approach. Further analysis with other data gave additional support and encouragement. While the approach was strictly empirical, the function did exhibit the correct form of response. After reading that science progresses by intuition, not just by cold logic, I was no longer embarrassed by the "guessing" approach. As often happens, progress with this effort opened up questions related to other factors such as nutrient levels and water availability. One of the parameters in the dynamic model depended on level of applied nutrients, the form of which suggested another mathematical function. In turn, one of its parameters depended on water availability. Now I was back to the level of the Mitscherlich equation from earlier times, but with a better function in hand. As discussed in Chapter 1, the hint for this improved function was identified as early as 1937 by Sir John Russell.

The next challenge was to expand the mathematical models to incorporate dry matter and plant nutrient accumulation as related to multiple levels of applied nutrients (N, P, K) in a mathematically self-consistent manner. The empirical model was expanded to a phenomenological model with a physical basis. This model was shown to apply to warm-season perennial grass for harvest intervals up to about 6 weeks, but failed for longer growth intervals and did not apply to annual grasses such as corn. Eventually, the growth model was modified again to cover these cases.

Many individuals deserve thanks for inspiration and help in my struggle to understand and describe science and engineering. Francis J. Hassler, J. van Schilfgaarde, Raymond J. Miller, and Ernest T. Smerdon nurtured the spirit of intellectual curiosity. Sigurd Melsted introduced me to the world of soil fertility and simple models. Thomas P. Smith, William G. Leseman, and John Dean encouraged research on water reclamation and reuse (which partially motivated the work on crop models). This led to many conversations with farmers Walter Vidak and J. L. Morgan about crop and animal management at the Tallahassee Southeast Farm. I have had the good fortune to work with several colleagues in agricultural research, including Stanley Wilkinson, Frank Martin, Eugene Kamprath, Gerald Evers, Donald Robinson, William Blue, Charles Ruelke, Fred Rhoads, Robert Stanley, John Moore, and George Hochmuth. The scope of field plot research by Glenn Burton at Tifton, Georgia, and William Adams at Watkinsville, Georgia, has been a particular source of inspiration. Special thanks are due Elizabeth Angley, who worked with me on the early mathematical models and developed computer programs for nonlinear regression. Marcus Allhands, Bill Reck, Denise Wilson, John Willcutts, Sherry Allick, Scott Thourot, and Kelly Brock worked on aspects of this subject during their student days. I especially thank my coauthor, Richard V. Scholtz III, for extensive discussions on mathematical modeling of crop systems as well as help with graphics in the text. Without his assistance, I could not have finished this project.

Allen R. Overman

Contents

MATHEMATICAL MODELS OF CROP GROWTH AND YIELD

1
Introduction

1.1 HISTORICAL BACKGROUND

A dramatic change occurred in human history after the last ice age just over ten thousand years ago. Mankind shifted from hunter/gatherers to the cultivation of plants and domestication of animals (Bronowski, 1973). The last four centuries have witnessed fundamental progress in the management of agriculture (McClelland, 1997; Schlebecker, 1975). Work on a particular machine, the reaper, has been described by Canine (1995). During this latter period rapid advances have been made in the field of science. The twentieth century has been characterized by unprecedented discoveries in physics by a blending of deduction and induction (Born, 1956; Lightman, 2000). Induction leads from specific observations to general laws, while deduction leads from general principles to specific predictions. Our approach in this book is founded on this blend. Ideas at the frontiers of science have been discussed by Bak (1996), Holton (1973), Lindley (2001), Pagels (1988), Penrose (1989), and Polkinghorne (1996).

Response of agricultural crops to management factors (such as available water, applied nutrients, plant density, harvest interval for perennials) has been of interest to farmers and agricultural scientists for a long time. Perhaps the most famous experiments in the world on crop production are those conducted at Rothamsted (England). John Bennet Lawes (1814–1900) devoted his family estate to the study and improvement of agricultural

1

production. An early account of this work can be found in Hall (1905). In 1834 Lawes initiated chemical experiments in a laboratory in his house. This was followed in 1837 by experiments with plants in pots. Then in 1840 he applied superphosphate to some of the fields on the farm. The results were so satisfactory that in 1842 he patented a process for manufacture of super-phosphate. In 1843 Joseph Henry Gilbert (1917–1901), who had taken his PhD with Justus Liebig, was employed as a chemist by Lawes. Their joint work with field plots and various crops continued over the next 57 years, leading to many publications in scientific journals. In 1919 Ronald Aylmer Fisher was hired as statistician at the Rothamsted Experimental Station, and over the next few years developed his analysis of variance and principles of experimental design (Box, 1978).

The approach to modeling in this book is very much motivated and guided by experimental results. Models have been developed and refined step by step rather than from some grand design, which then fills in the details. And while considerable progress has been made, this is very much a picture of a process which is still evolving. In this work we have not attempted to review the broad field of crop modeling, but rather have focused on models which we have found useful in practice. The reader will undoubtedly detect our preference for analytical functions, in contrast to finite difference procedures (Ford, 1999). We have never found finite difference to be an easy alternative to analytical functions. Students should still learn calculus, differential equations, and physics.

1.2   YIELD RESPONSE MODELS

One of the earliest efforts at modeling crop response to management factors was that of E. A. Mitscherlich, which stimulated a lot of interest and generated considerable controversy (Russell, 1912; van der Paauw, 1952). The Mitscherlich equation can be written as

$$Y = Y_0 + (Y_m - Y_0)[1 - \exp(-cN)] \qquad [1.1]$$

where N = applied nutrient; Y = dry matter yield; Y_0 = dry matter yield at $N = 0$; Y_m = maximum dry matter yield at high N; c = nutrient response coefficient. Yield values are all assumed to be positive. Equation [1.1] can be rearranged to the alternate form

$$Y_m - Y = (Y_m - Y_0)\exp(-cN) \qquad [1.2]$$

Equation [1.2] suggests plotting $(Y_m - Y)$ vs. N on semilog paper, which requires an estimate of Y_m. If this plot follows a straight line, then linear regression can be performed on $\ln(Y_m - Y)$ vs. N to obtain estimates for

$(Y_m - Y_0)$ and c. Figure 1.1 shows data from Mitscherlich (1909), as reported by Russell (1912, p. 25), for response of oats (*Avena sativa* L.) in pots to applications of P_2O_5. The semilog plot is shown in Fig. 1.2, where the line is given by

$$\ln(61.0 - Y) = \ln(Y_m - Y_0) - c(P_2O_5) = 3.94 - 4.16P_2O_5 \qquad r = -0.9939$$
$$[1.3]$$

with a correlation coefficient of -0.9939. It follows that $(Y_m - Y_0) = 51.4\,\text{g}$ and that $Y_0 = 9.6\,\text{g}$. The line in Fig. 1.2 is drawn from Eq. [1.3] and the curve in Fig. 1.1 from

$$Y = 9.6 + 51.4[1 - \exp(-4.16P_2O_5)] \qquad [1.4]$$

The fit of the Mitscherlich equation to these data appears quite reasonable.

It turns out that even earlier data do not conform as well to the Mitscherlich equation. Data from Hellriegel and Wilfarth (1888), as reported by Russell (1912, p. 32), for barley (*Hordeum vulgare* L.) in sand cultures exhibit the trend shown in Fig. 1.3. As pointed out by Russell (1937, p. 135) these data follow an S-shaped curve. We now refer to this form as sigmoid. To illustrate the problem with the Mitscherlich model, the hori-

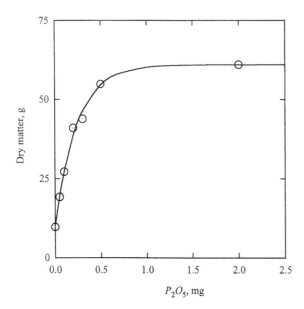

Figure 1.1 Yield response of oats to applied phosphorus. Data from Mitscherlich (1909) as reported by Russell (1912). Curve drawn from Eq. [1.4].

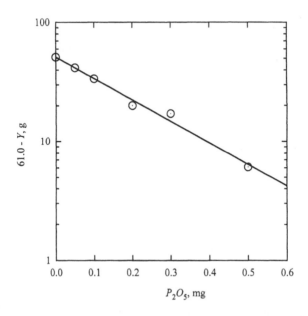

Figure 1.2   Semilog plot of yield response of oats to applied phosphorus for experiment of Mitscherlich (1909). Line drawn from Eq. [1.3].

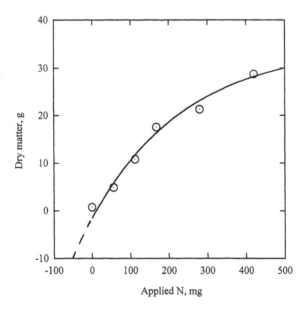

Figure 1.3   Yield response of barley to applied nitrogen. Data from Hellriegel and Wilfarth (1888) as reported by Russell (1912). Curve drawn from Eq. [1.6].

zontal and vertical axes have been expanded to include negative values. Figure 1.4 shows the semilog plot, with the line given by

$$\ln(35.0 - Y) = 3.60 - 0.00400N \qquad r = -0.9905 \qquad [1.5]$$

where N = applied N, mg. This leads to $(Y_m - Y_0) = 36.6$ g and to $Y_0 = -1.6$ g. The curve in Fig. 1.3 is drawn from

$$Y = -1.6 + 36.6[1 - \exp(-0.00400N)] \qquad [1.6]$$

While negative values of N might signify reduction of soil N below background level, negative dry matter yield has no physical meaning. This dilemma has generally been ignored, but an alternative is to seek a better mathematical model.

The logistic equation is proposed as an alternate model. It is given by

$$Y = \frac{A}{1 + \exp(b - cN)} \qquad [1.7]$$

where Y = dry matter yield; N = applied nutrient; A = maximum dry matter yield at high N; b = intercept parameter; c = nutrient response coefficient. This model exhibits the sigmoid shape suggested by Russell (1937; 1950, p. 58), and Y remains positive for all values of N. The data of

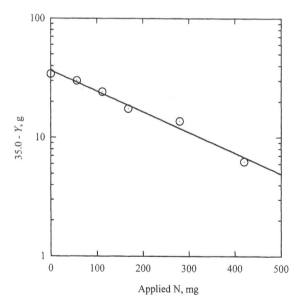

Figure 1.4 Semilog plot of yield response of barley to applied nitrogen for experiment of Hellriegel and Wilfarth (1888). Line drawn from Eq. [1.5].

Hellriegel and Wilfarth are shown again in Fig. 1.5, where the curve is drawn from

$$Y = \frac{26.0}{1 + \exp(2.30 - 0.0160N)} \qquad r = 0.9804 \qquad \text{[1.8]}$$

Parameters $A, b,$ and c were selected by visual inspection. The method of nonlinear regression for parameter estimation will be discussed later. It can be shown that the logistic equation asymptotically approaches A for large N and approaches zero for reduced (negative) N. Intercept parameter b relates to the reference state ($N = 0$), while c controls the rate of rise of Y on N. Furthermore, for $N = N_{1/2} = b/c$, it can be shown that $Y = A/2$.

Russell (1912, p. 43) also presents data from Hellriegel et al. (1898) for response of barley to applied K_2O, as shown in Fig. 1.6. In this case the curve is drawn from

$$Y = \frac{36.5}{1 + \exp(2.40 - 0.0200K_2)} \qquad r = 0.9984 \qquad \text{[1.9]}$$

The sigmoid relationship is clearly evident.

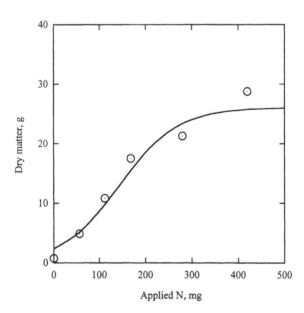

Figure 1.5 Yield response of barley to applied nitrogen. Data from Hellriegel and Wilfarth (1888) as reported by Russell (1912). Curve drawn from Eq. [1.8].

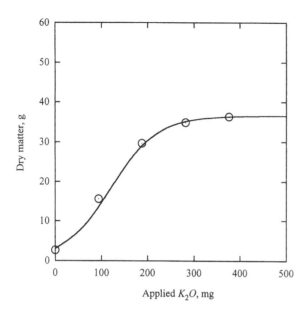

Figure 1.6 Yield response of barley to applied potassium. Data from Hellriegel et al. (1898) as reported by Russell (1912). Curve drawn from Eq. [1.9].

Another application of the logistic model is shown in Fig. 1.7 for response of wheat (*Triticum aestivum* L.) to applied nitrogen at Rothamsted for the period 1852 through 1864 (Russell, 1912, p. 34). The soil is a heavy loam underlain by chalk (Hall, 1905, p. 24). Curves are drawn from

$$Y = \frac{A}{1 + \exp(0.50 - 0.0200N)} \qquad [1.10]$$

where $A = 2.75, 5.90$, and 8.65 Mg ha^{-1} for grain, straw, and total dry matter, respectively. According to this analysis grain constitutes a fixed fraction of total dry matter at all applied N levels.

It appears that the logistic equation can be used to describe response of crops to applied N, P, and K. Further development and application of the model are discussed in Chapters 2 and 4.

1.3 GROWTH MODELS

Examples given above focused on response of seasonal dry matter to applied nutrients. We now turn attention to accumulation of dry matter and plant

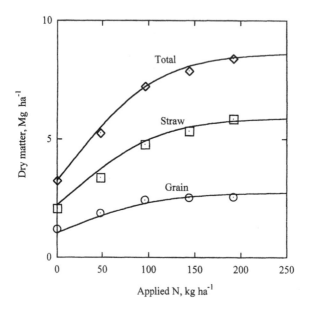

Figure 1.7 Yield response of wheat to applied nitrogen. Data from Russell (1912). Curves drawn from Eq. [1.10] with $A = 2.75, 5.90$, and 8.65 Mg ha^{-1} for grain, straw, and total, respectively.

nutrients with time. Overman (1984) described crop growth with the probability model

$$Y = \frac{A}{2}\left[1 + \text{erf}\left(\frac{t - \mu}{\sqrt{2}\sigma}\right)\right] \qquad [1.11]$$

where Y = accumulated dry matter, Mg ha^{-1}; t = calendar time since Jan. 1, wk; A = maximum accumulated dry matter, Mg ha^{-1}; μ = time to the mean of the dry matter distribution, wk; σ = time spread of the dry matter distribution, wk. The error function in Eq. [1.11] is defined by

$$\text{erf}\, x = \frac{2}{\sqrt{\pi}} \int_0^x \exp(-u^2)\, du \qquad [1.12]$$

which can be evaluated from mathematical tables (Abramowitz and Stegun, 1965). The model can also be normalized by defining

$$F = \frac{Y}{A} = \frac{1}{2}\left[1 + \text{erf}\left(\frac{t - \mu}{\sqrt{2}\sigma}\right)\right] \qquad [1.13]$$

Equation [1.13] produces a straight line on probability paper. Estimates of μ and σ can be obtained by graphical means from this plot of data.

Application of this simple growth model is now demonstrated. Leukel et al. (1934) conducted a field study with bahiagrass (*Paspalum notatum* Flügge) on a sandy soil in Florida. Applied N level was 295 kg ha^{-1} on the fertilized plots. Data for dry matter and plant N distribution over the season with fertilizer and irrigation are given in Table 1.1 and without either fertilizer or irrigation in Table 1.2. Accumulation of dry matter with time is shown in Fig. 1.8, with the normalized yield distribution shown in Fig. 1.9. The parameters are estimated by graphical means to be $\mu = t(F = 50\%) = 29.5$ wk and $\sigma = [t(F = 84\%) - t(F = 16\%)]/2 = [36.5 - 23.5]/2 = 6.5$ wk. It follows that the line in Fig. 1.9 is drawn from

Table 1.1 Dry matter and plant N accumulation with time for bahiagrass grown with fertilizer and irrigation at Gainesville, FL

t wk	ΔY Mg ha^{-1}	Y Mg ha^{-1}	F_y	ΔN_u kg ha^{-1}	N_u kg ha^{-1}	F_n	N_c g kg^{-1}
—		0	0		0	0	
	0.061			0.90			14.8
20.3		0.061	0.029		0.90	0.024	
	0.065			1.05			16.2
22.3		0.126	0.060		1.95	0.052	
	0.311			7.18			23.1
24.4		0.437	0.208		9.13	0.244	
	0.414			6.78			16.4
27.3		0.851	0.405		15.91	0.426	
	0.194			3.04			15.7
29.4		1.045	0.498		18.95	0.507	
	0.227			3.98			17.5
31.4		1.272	0.606		22.93	0.613	
	0.174			3.23			18.6
34.3		1.446	0.689		26.16	0.699	
	0.331			5.16			15.6
36.7		1.777	0.846		31.32	0.838	
	0.185			3.50			18.9
39.6		1.962	0.934		34.82	0.932	
	0.138			2.56			18.6
43.7		2.100	1		37.38	1	

Data for dry matter and plant N adapted from Leukel et al. (1934, table 1). t = calendar time since Jan. 1, wk; Y = dry matter yield, Mg ha^{-1}; N_u = plant N uptake, kg ha^{-1}; N_c = plant N concentration, g kg^{-1}.

Table 1.2 Dry matter and plant N accumulation with time for bahiagrass grown without either fertilizer or irrigation at Gainesville, FL

t wk	ΔY Mg ha^{-1}	Y Mg ha^{-1}	F_y	ΔN_u kg ha^{-1}	N_u kg ha^{-1}	F_n	N_c g kg^{-1}
—		0	0		0	0	
	0.086			1.18			13.7
20.3		0.086	0.086		1.18	0.076	
	0.045			0.68			15.1
22.3		0.131	0.132		1.86	0.120	
	0.144			2.23			15.5
24.4		0.275	0.276		4.09	0.264	
	0.192			3.08			16.0
27.3		0.467	0.469		7.17	0.463	
	0.069			1.06			15.4
29.4		0.536	0.539		8.23	0.532	
	0.124			2.12			17.1
31.4		0.660	0.663		10.35	0.669	
	0.138			2.26			16.4
34.3		0.798	0.802		12.61	0.815	
	0.077			1.00			13.0
36.7		0.875	0.879		13.61	0.879	
	0.079			1.30			16.5
39.6		0.954	0.959		14.91	0.963	
	0.041			0.57			13.9
43.7		0.995	1		15.48	1	

Data for dry matter and plant N adapted from Leukel et al. (1934, table 1). t = calendar time since Jan. 1, wk; Y = dry matter yield, Mg ha^{-1}; N_u = plant N uptake, kg ha^{-1}; N_c = plant N concentration, g kg^{-1}.

$$F_y = \frac{1}{2}\left[1 + \mathrm{erf}\left(\frac{t - 29.5}{9.20}\right)\right] \qquad [1.14]$$

and the curve in Fig. 1.8 is drawn from

$$Y = \frac{A}{2}\left[1 + \mathrm{erf}\left(\frac{t - 29.5}{9.20}\right)\right] \qquad [1.15]$$

where $A = 2.100$ Mg ha^{-1} with fertilizer and irrigation, and $A = 0.955$ Mg ha^{-1} without either fertilizer or irrigation. Plant N distribution is also listed in Tables 1.1 and 1.2. These data are used to show the dependence of normalized plant N upon normalized dry matter (F_n vs. F_y) in Fig. 1.10, where the line is given by

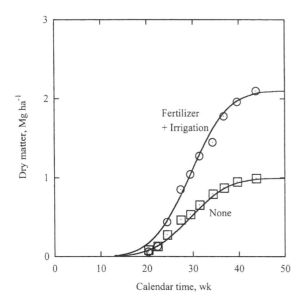

Figure 1.8 Dry matter accumulation with time for bahiagrass. Data from Leukel et al. (1934). Curves drawn from Eq. [1.15] with $A = 2.100$ Mg ha^{-1} for fertilizer and irrigation and $A = 0.995$ Mg ha^{-1} without either fertilizer or irrigation.

$$F_n = -0.001 + 1.006F_y \quad r = 0.9992 \qquad [1.16]$$

Average plant N concentration is estimated from the ratio of plant N uptake to dry matter and is equal to $N_c = N_u/Y = 37.38/2.100 = 17.8$ g kg^{-1} with fertilizer and irrigation, and 15.6 g kg^{-1} without either fertilizer or irrigation. It appears that the simple probability model describes the data reasonably well. It is strictly a regression model without any physical basis at this point.

A growth model published by Overman (1998) is now used to simulate dry matter accumulation with time. Dry matter accumulation is given by

$$Y = AQ \qquad [1.17]$$

where Y = accumulated dry matter; A = yield factor; Q = growth quantifier. The growth quantifier is defined by

$$Q = \exp(\sqrt{2}\sigma c x_i)\{(1 - kx_i)(\text{erf}\,x - \text{erf}\,x_i) - \frac{k}{\sqrt{\pi}}[\exp(-x^2) - \exp(-x_i^2)]\} \qquad [1.18]$$

where the error function is defined by

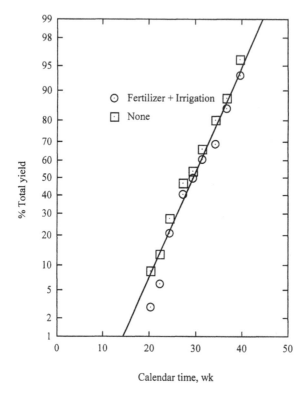

Figure 1.9 Probability plot of dry matter accumulation with time for bahiagrass for experiment of Leukel et al. (1934). Line drawn from Eq. [1.14].

$$\operatorname{erf} x = \frac{2}{\sqrt{\pi}} \int_0^x \exp(-u^2) \, du \qquad\qquad [1.19]$$

and the dimensionless time variable x is defined by

$$x = \frac{t - \mu}{\sqrt{2}\sigma} + \frac{\sqrt{2}\sigma c}{2} \qquad\qquad [1.20]$$

where t = calendar time since Jan. 1, wk; μ = mean time of the environmental function, wk; σ = time spread of the environmental function, wk; c = exponential coefficient in the intrinsic growth function, wk^{-1}; k = curvature factor in the intrinsic growth function. In Eq. [1.18], x_i relates to time of initiation of significant growth, t_i. Values of erf x can be obtained from mathematical tables (Abramowitz and Stegun, 1965), recalling that $\operatorname{erf}(-x) = \operatorname{erf}(+x)$. This model will be discussed in more detail in Chapters 3 and 4.

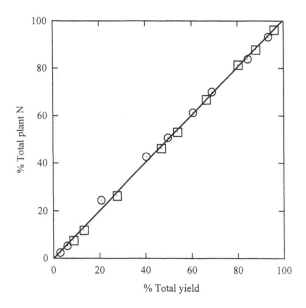

Figure 1.10 Dependence of normalized plant N accumulation on normalized dry matter accumulation by bahiagrass for the experiment of Leukel et al. (1934). Line drawn from Eq. [1.16].

Application of the growth model is now illustrated with data from the literature. Knowles and Watkins (1931) conducted a field experiment in 1930 with wheat on heavy clay soil at Essex, England. Data have been summarized by Russell (1950, p. 29) and are given in Table 1.3. Dry matter accumulation is shown in Fig. 1.11. Values were recorded for a fixed number of tillers, without specifying area involved. To calibrate the model we choose the following parameters: $t_i = 16$ wk, $\mu = 26$ wk, $\sqrt{2}\sigma = 8$ wk, $c = 0.2$ wk^{-1}, $k = 5$. Then Eq. [1.20] becomes

$$x = \frac{t - 26}{8} + 0.8 = \frac{t - 19.6}{8} \qquad [1.21]$$

and Eq. [1.18] becomes

$$Q = 0.487\{3.25(\mathrm{erf}\,x + 0.475) - 2.821[\exp(-x^2) - 0.817]\} \qquad [1.22]$$

Computations from Eqs. [1.21] and [1.22] are listed in Table 1.4. The yield factor A is estimated by matching the function to yield at 26 wk, so that

$$Y = \left(\frac{12.00}{2.32}\right)Q = 5.17Q \qquad [1.23]$$

Table 1.3 Dry matter and plant nutrient accumulation by wheat at Essex, England

Time wk	Dry matter kg	Plant N g	Plant P g	Plant K g	Plant Ca g
17.3	0.77	27	3.2	26	5.1
20.3	2.97	55	10.5	79	12.8
22.3	6.19	80	18.0	140	21.6
24.3	9.51	90	24.4	149	25.6
26.3	12.27	96	24.8	121	26.8
28.3	14.35	110	27.0	103	26.5
29.3	15.06	109	27.8	94	22.8
30.3	14.73	109	27.8	86	22.0
31.3	14.21	109	27.8	75	21.6

Data adapted from Russell (1950, p. 29).

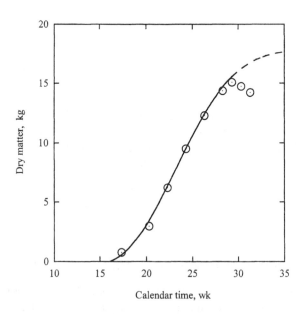

Figure 1.11 Dry matter accumulation with time for wheat. Data from Knowles and Watkins (1931) as summarized by Russell (1950). Curve drawn from Eqs. [1.21] through [1.23].

Table 1.4 Model calculations for wheat grown at Essex, England

t wk	x	erf$x$	$\exp(-x^2)$	Q	Y kg
16	−0.450	−0.475	0.817	0	0
17	−0.325	−0.354	0.900	0.077	0.40
18	−0.200	−0.223	0.961	0.201	1.04
19	−0.075	−0.085	0.9944	0.374	1.93
20	0.050	0.056	0.9975	0.592	3.06
21	0.175	0.195	0.970	0.850	4.40
22	0.300	0.329	0.914	1.14	5.89
23	0.425	0.451	0.835	1.44	7.45
24	0.550	0.563	0.739	1.75	9.05
25	0.675	0.660	0.634	2.05	10.59
26	0.800	0.742	0.527	2.32	12.00
27	0.925	0.809	0.425	2.57	13.29
28	1.050	0.862	0.332	2.78	14.39
29	1.175	0.903	0.251	2.96	15.30
30	1.300	0.934	0.185	3.10	16.02
32	1.550	0.972	0.090	3.29	17.00
34	1.800	0.989	0.039	3.39	17.51

Estimates of dry matter accumulation in Table 1.4 are made from Eq. [1.23], which is used to calculate the curve in Fig. 1.11. The decrease in dry matter toward the end of the experiment could have been a result of lodging caused by wet and windy conditions which occurred around mid-July (Knowles and Watkins, 1931).

We next focus on plant nutrient accumulation with time. One possibility is to simulate this variable with time similar to dry matter accumulation. A better alternative appears to be to use a phase plot of plant nutrient accumulation vs. dry matter accumulation for each element, and then combine these results with dry matter simulation. We assume the appropriate phase relation to be given by the hyperbolic equation

$$N_u = \frac{N_{um} Y}{K_y + Y} \qquad [1.24]$$

where N_u = plant nutrient accumulation; N_{um} = potential maximum plant nutrient accumulation; K_y = yield response coefficient. Note that for $Y = K_y$, $N_u = N_{um}/2$. It is easily shown that Eq. [1.24] can be converted to the linear form

$$\frac{Y}{N_u} = \frac{K_y}{N_{um}} + \frac{Y}{N_{um}} \tag{1.25}$$

Figure 1.12 shows the phase plot for plant N. Linear regression of Y/N_u vs. Y leads to

$$\frac{Y}{N_u} = 0.031 + 0.00728\,Y \qquad r = 0.9917 \tag{1.26}$$

It follows that the hyperbolic phase relation for plant N becomes

$$N_u = \frac{137\,Y}{4.1 + Y} \tag{1.27}$$

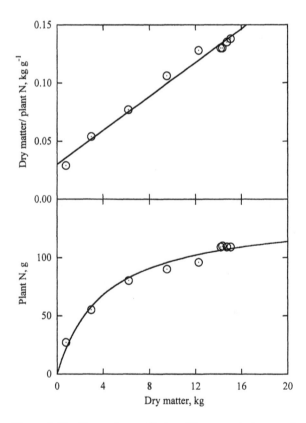

Figure 1.12 Dependence of plant N accumulation on dry matter accumulation by wheat for experiment of Knowles and Watkins (1931). Line drawn from Eq. [1.26] and curve drawn from Eq. [1.27].

The line and curve in Fig. 1.12 are drawn from Eqs. [1.26] and [1.27], respectively. Estimates of plant N accumulation are calculated by substituting estimates of dry matter accumulation from Table 1.4 into Eq. [1.27]. Results are shown in Fig. 1.13, where agreement between simulation and data appears acceptable. A similar procedure for plant P accumulation leads to the linear relation

$$\frac{Y}{P_u} = 0.217 + 0.0213\,Y \qquad r = 0.9936 \qquad\qquad [1.28]$$

and the hyperbolic phase relation

$$P_u = \frac{47.0\,Y}{10.2 + Y} \qquad\qquad [1.29]$$

which are shown in Fig. 1.14. Estimates of plant P accumulation are calculated by substituting estimates of dry matter accumulation from Table 1.4 into Eq. [1.29], with results shown in Fig. 1.15. Excellent agreement between simulation and data is apparent. Analysis of plant K becomes somewhat more complicated, as shown in Fig. 1.16. Data in Table 1.3 show that plant K accumulation actually decreased after 24 wk. This effect may be noted in

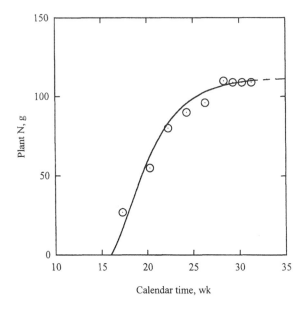

Figure 1.13 Plant N accumulation with time for wheat. Data from Knowles and Watkins (1934). Curve drawn from Eqs. [1.21] through [1.23] and [1.27].

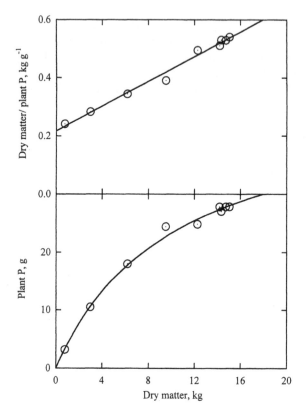

Figure 1.14 Dependence of plant P accumulation on dry matter accumulation by wheat for experiment of Knowles and Watkins (1931). Line drawn from Eq. [1.28] and curve drawn from Eq. [1.29].

the phase plot. As a result only the first three data points in Table 1.3 are used to obtain

$$\frac{Y}{K_u} = 0.0284 + 0.00265\,Y \qquad r = 0.9867 \tag{1.30}$$

with the resulting hyperbolic equation

$$K_u = \frac{377\,Y}{10.7 + Y} \tag{1.31}$$

Estimates of plant K accumulation are calculated by substituting estimates of dry matter accumulation from Table 1.4 into Eq. [1.31], with results shown in Fig. 1.17. The simulation only holds up to 23.5 wk, at which

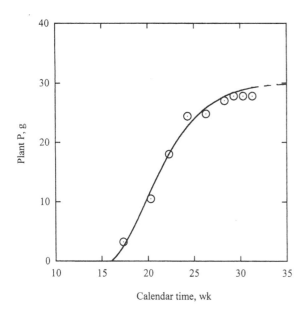

Figure 1.15 Plant P accumulation with time for wheat. Data from Knowles and Watkins (1934). Curve drawn from Eqs. [1.21] through [1.23] and [1.29].

time the loss of plant K apparently begins. After that time, plant K appears to follow exponential decay given by

$$K_u = 160 \exp[-0.094(t - 23.5)] \tag{1.32}$$

This phenomenon will be discussed in more detail in Section 4.6. Finally, results for plant Ca accumulation are shown in Fig. 1.18, where the line is given by

$$\frac{Y}{Ca_u} = 0.131 + 0.0272Y \qquad r = 0.9934 \tag{1.33}$$

with the curve drawn from the hyperbolic equation

$$Ca_u = \frac{36.8Y}{4.8 + Y} \tag{1.34}$$

The last three data points were omitted from regression analysis. Estimates of plant Ca accumulation are calculated by substituting estimates of dry matter accumulation from Table 1.4 into Eq. [1.34], with results shown in Fig. 1.19. Simulation is quite good up until 28 weeks when the plants begin to lose calcium.

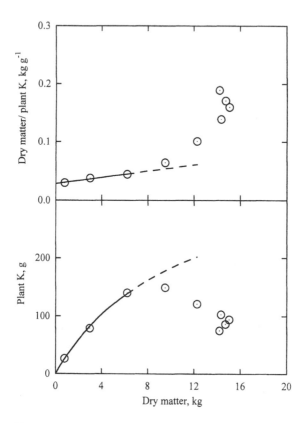

Figure 1.16   Dependence of plant K accumulation on dry matter accumulation by wheat for experiment of Knowles and Watkins (1931). Line drawn from Eq. [1.30] and curve drawn from Eq. [1.31].

1.4   ENVIRONMENTAL INPUT

There is clearly an environmental, or climatic, input to crop growth. This is evident in seasonal changes in rates of growth caused by variations in solar radiation over the calendar year. In this discussion we focus on the northern hemisphere. Data for two climatic variables, air temperature and sunshine, are summarized for Rothamsted Experiment Station (Hall, 1905, table VI) in Table 1.5. Average temperature data for 26 years (1878–1903) and average sunshine data for 11 years (1892, 1893, 1895–1903) are shown in Fig. 1.20. Both variables appear to follow somewhat Gaussian distributions over the year. To test this idea, normalized values of both variables are calculated

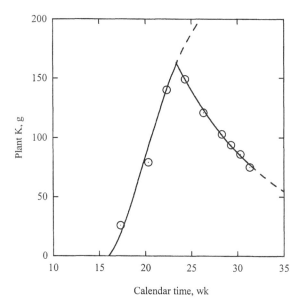

Figure 1.17 Plant K accumulation with time for wheat. Data from Knowles and Watkins (1934). Curves drawn from Eqs. [1.21] through [1.23], [1.31], and [1.32].

in Table 1.5 and are shown in Fig. 1.21. To improve the analyses, base values are subtracted from each variable. Since both distributions essentially follow straight lines, the idea seems reasonable as a first approximation at least. By visual inspection we obtain $\mu = 25.5$ wk and $\sigma = 10.5$ wk for sunshine, and $\mu = 28.5$ wk and $\sigma = 10.5$ wk for temperature. The lines are drawn from

$$F = \frac{1}{2}\left[1 + \operatorname{erf}\left(\frac{t - \mu}{\sqrt{2}\sigma}\right)\right]$$

[1.35]

with parameters given above. Another measure of the environmental input is listed in Table 1.6 and shown in Fig. 1.22 for average solar radiation during the period 1931–1940 at Rothamsted, England (Russell, 1950, table 75). Parameters are estimated to be $\mu = 25.0$ wk and $\sigma = 10.5$ wk from Fig. 1.23, with the line drawn from Eq. [1.35].

Incorporation of an environmental function into the growth models will be discussed in Chapter 3.

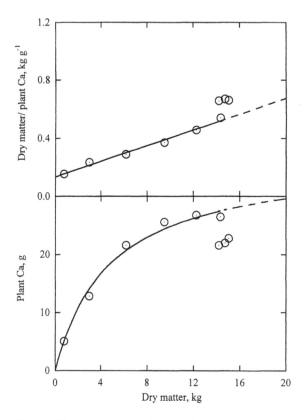

Figure 1.18 Dependence of plant Ca accumulation on dry matter accumulation by wheat for experiment of Knowles and Watkins (1931). Line drawn from Eq. [1.33] and curve drawn from Eq. [1.34].

1.5 SUMMARY

In this chapter we have used data from the literature to illustrate the application of several mathematical models. Some of these data came from studies as early as the mid-1800s. The purpose has been to show that some of the key patterns in the data have been present for a long time. One criticism of the approach here is that it is just "curve fitting." In succeeding chapters we hope to dispel this claim. Such a claim in science is not new. When Niels Bohr published his model of the atom in 1913, it was labeled "number juggling" (Cline, 1987, p. 106). Acknowledging that achievement of the ideal model for crop simulation seems rather illusive,

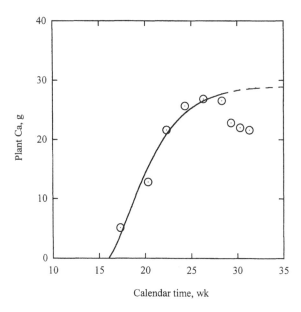

Figure 1.19 Plant Ca accumulation with time for wheat. Data from Knowles and Watkins (1934). Curve drawn from Eqs. [1.21] through [1.23] and [1.34].

we will adopt the acronym attributed to physicist John Bell: FAPP = for all practical purposes (Wick, 1995). This is also stated as the "first approximation method." We subscribe to the view stated by Richard Feynman that mathematics is not just a language, but is also a tool for reasoning (Feynman, 1965, p. 40).

This chapter has been somewhat of an introduction to several models. Succeeding chapters will deal with these models at increasingly complex levels. The idea is to develop the motivation for learning the math and physics concepts by demonstrating the practical application of the models. A brief review of some of the models covered in this book has been presented by Overman and Wilson (1999).

EXERCISES

1.1 For the data of Mitscherlich given in the table below
    a. Construct the plot shown in Fig. 1.1 for yourself.
    b. For each level of applied P_2O_5, add the column $61.0 - Y$ to the table.

Table 1.5 Temperature and sunshine measurements at Rothamsted, England

t wk	T °F	$T-35$ °F	$\sum(T-35)$ °F	F_t	S h mo^{-1}	$S-25$ h mo^{-1}	$\sum(S-25)$ h mo^{-1}	F_s
0.0			0	0			0	0
	36.6	1.6			46.4	21.4		
4.4			1.6	0.010			21.4	0.017
	38.2	3.2			69.2	44.2		
8.6			4.8	0.031			65.6	0.051
	40.9	5.9			114.6	89.6		
13.0			10.7	0.069			155.2	0.121
	45.5	10.5			170.3	145.3		
17.3			21.2	0.137			300.5	0.234
	51.2	16.2			199.9	174.9		
21.7			37.4	0.242			475.4	0.370
	57.5	22.5			201.9	176.9		
26.0			59.9	0.387			652.3	0.507
	60.7	25.7			217.5	192.5		
30.4			85.6	0.553			844.8	0.657
	59.9	24.9			201.9	176.9		
34.9			110.5	0.714			1020.9	0.794
	55.9	20.9			158.3	133.3		
39.1			131.4	0.849			1154.2	0.898
	48.0	13.0			106.1	81.1		
43.6			144.4	0.933			1235.3	0.961
	42.6	7.6			57.0	32.0		
47.9			152.0	0.983			1267.3	0.986
	37.7	2.7			43.2	18.2		
52.3			154.7	1			1285.5	1

Temperature and sunshine data adapted from Hall (1905).
t = calendar time since Jan. 1, wk; T = average air temperature, °F; S = average hours of sunshine per month, h mo^{-1}.

c. Construct the plot shown in Fig. 1.2 for yourself.
d. Perform linear regression on $\ln(61.0 - Y)$ vs. P_2O_5 to obtain the parameters $\ln(Y_m - Y_0)$ and c in Eq. [1.3].
e. Draw the regression line on Fig. 1.2.
f. Draw the regression curve on Fig. 1.1.
g. Discuss the results.

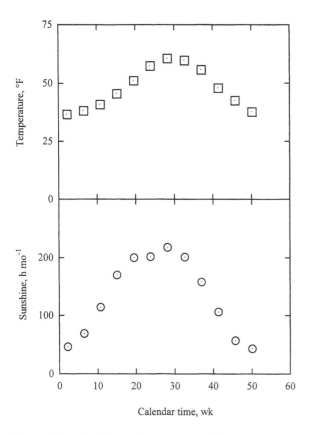

Figure 1.20 Variation of average monthly temperature and sunshine with calendar time for Rothamsted, England. Data from Hall (1905).

Dry matter response of oats to applied P_2O_5
for Mitscherlich (1909) experiment in Germany

P_2O_5 mg	Dry matter g
0.00	9.8
0.05	19.3
0.10	27.2
0.20	41.0
0.30	43.9
0.50	54.9
2.00	61.0

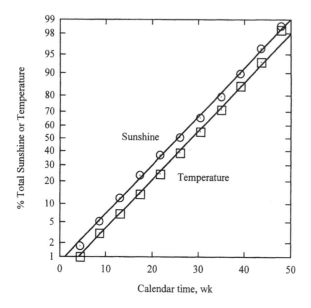

Figure 1.21 Probability plot of average monthly temperature and sunshine for Rothamsted, England for data of Hall (1905). The lines are drawn from Eq. [1.35] with $\mu = 28.5$ wk and $\sigma = 10.5$ wk for temperature and $\mu = 25.5$ wk and $\sigma = 10.5$ wk for sunshine.

1.2 For the data of Hellriegel and Wilfarth given in the table below
   a. Construct the plot shown in Fig. 1.3.
   b. For each level of applied N, add the column $35.0 - Y$ to the table.
   c. Construct the plot shown in Fig. 1.4.
   d. Perform linear regression on $\ln(35.0 - Y)$ vs. N to obtain the parameters $\ln(Y_m - Y_0)$ and c in Eq. [1.5].
   e. Draw the regression line on Fig. 1.4.
   f. Draw the regression curve on Fig. 1.3.
   g. Construct the plot shown in Fig. 1.5.
   h. For each level of applied N, add the column $26.0/Y - 1)$ to the table.
   i. Perform linear regression on $\ln(26.0/Y - 1)$ vs. N to obtain parameters b and c in Eq. [1.7].
   j. Use these parameters to construct the regression curve for Fig. 1.5.
   k. Discuss the results.

Table 1.6 Solar radiation measurements at Rothamsted, England

t wk	R kcal cm^{-2} mo^{-1}	$\sum R$ kcal cm^{-2} mo^{-1}	F
0.0		0	0
	1.68		
4.4		1.68	0.022
	2.73		
8.6		4.41	0.058
	6.28		
13.0		10.69	0.140
	7.82		
17.3		18.51	0.243
	10.95		
21.7		29.46	0.386
	12.16		
26.0		41.62	0.545
	11.20		
30.4		52.82	0.692
	9.49		
34.9		62.31	0.817
	6.83		
39.1		69.14	0.906
	4.01		
43.6		73.15	0.959
	1.91		
47.9		75.06	0.984
	1.24		
52.3		76.3	1

Solar radiation data adapted from Russell (1950).
t = calendar time since Jan. 1, wk; R = average monthly solar radiation, kcal cm^{-2} mo^{-1}.

Dry matter response of barley to applied N for
Hellriegel and Wilfarth (1888) experiment in
Germany

N mg	Dry matter g
0	0.74
56	4.86
112	10.80
168	17.53
280	21.29
420	28.73

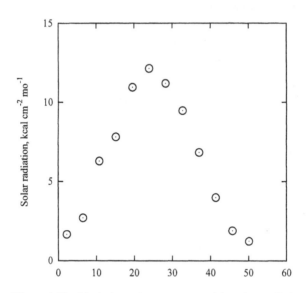

Figure 1.22 Variation of average monthly solar radiation with calendar time for Rothamsted, England. Data from Russell (1950).

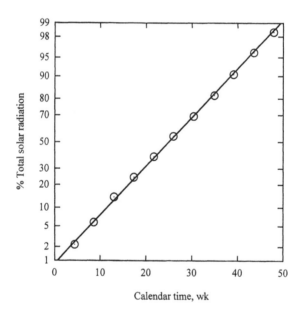

Figure 1.23 Probability plot of average monthly solar radiation for Rothamsted, England for data of Russell (1950). The line is drawn from Eq. [1.35] with $\mu = 25.0$ wk and $\sigma = 10.5$ wk.

1.3 For the data of Hellriegel et al. given below
   a. Construct the plot shown in Fig. 1.6.
   b. For each level of applied K_2O, add the column $36.5/Y - 1$ to the table.
   c. Perform linear regression on $\ln(36.5/Y - 1)$ vs. K_2O to obtain the parameters b and c in Eq. [1.9].
   d. Use these parameters to construct the regression curve for Fig. 1.6.
   e. Discuss the results.

Dry matter response of barley to applied K_2O for Hellriegel et al. (1898) experiment in Germany

K_2O mg	Dry matter g
0	2.66
94	15.6
188	29.7
282	34.9
376	36.3

1.4 For the data of Russell given in the table below
   a. Construct the plot shown in Fig. 1.7.
   b. For each level of applied N, add the columns $A/Y - 1$ to the table for $A = 2.75$. $5.90, 8.65\,\mathrm{Mg\ ha^{-1}}$ for grain, straw, and total dry matter, respectively.
   c. Perform linear regression on $\ln(8.65/Y - 1)$ vs. N to obtain parameters b and c in Eq. [1.10] for total dry matter.
   d. Use these parameters to construct the regression curves in Fig. 1.7.
   e. Discuss the results.

Dry matter response of wheat to applied N for experiment at Rothamsted, England (Russell, (1912)

N kg ha^{-1}	Dry matter, Mg ha^{-1}		
	Grain	Straw	Total
0	1.19	2.08	3.27
48	1.88	3.39	5.27
96	2.44	4.79	7.23
144	2.54	5.35	7.89
192	2.56	5.85	8.41

1.5 Data of Leukel et al. (1934) are given in the table below.

Dry matter and plant N accumulation with time for bahiagrass grown with fertilizer ($N = 295\,kg\ ha^{-1}$) and irrigation at Gainesville, FL

t wk	ΔY Mg ha^{-1}	Y Mg ha^{-1}	F_y	ΔN_u kg ha^{-1}	N_u kg ha^{-1}	F_n	N_c g kg^{-1}
—		0	0		0	0	
	0.390			11.7			30.0
20.7		0.390	0.073		11.7	0.092	
	0.538			9.7			18.0
22.6		0.928	0.174		21.4	0.169	
	0.801			18.4			23.0
25.1		1.729	0.324		39.8	0.315	
	0.310			7.1			22.9
27.6		2.039	0.382		46.9	0.371	
	0.734			15.5			21.1
30.1		2.773	0.520		62.4	0.493	
	0.501			12.1			24.2
31.6		3.274	0.614		74.5	0.589	
	0.458			10.9			23.8
33.1		3.732	0.700		85.4	0.675	
	0.400			10.9			27.3
35.6		4.132	0.775		96.3	0.761	
	0.494			11.7			23.7
37.6		4.626	0.867		108.0	0.854	
	0.420			10.7			25.5
39.6		5.046	0.946		118.7	0.938	
	0.144			3.7			25.7
41.6		5.190	0.973		122.4	0.968	
	0.063			1.6			25.4
43.6		5.253	0.985		124.0	0.980	
	0.081			2.5			30.9
45.6		5.334	1		126.5	1	

Data for dry matter and plant N adapted from Leukel et al. (1934, Table 5). t = calendar time since Jan. 1, wk; Y = accumulated dry matter, Mg ha^{-1}; N_u = accumulated plant N uptake, kg ha^{-1}; N_c = plant N concentration, g kg^{-1}.

a. Plot cumulative sum of dry matter and plant N uptake vs. time on linear graph paper.
b. Plot normalized dry matter and plant N uptake vs. time on probability graph paper.

c. Estimate parameters μ and σ by visual inspection from the probability graph.
d. Draw the curves on (a).
e. Plot normalized plant N uptake vs. normalized dry matter on linear graph paper.
f. Discuss the results.

1.6 For the data of Russell (1950) given in Table 1.6 for solar radiation at Rothamsted, England
a. Calculate the cumulative sum of adjusted solar radiation, $R - 5$, with time.
b. Calculate the normalized adjusted solar radiation with time.
c. Plot (b) on probability paper.
d. Estimate time to mean μ and standard deviation σ from the graph.
e. Compare these results to those for the unadjusted solar radiation.
f. Discuss use of temperature, sunlight, and solar radiation as an environmental function for the growth model.

1.7 Consider data for corn given in the table below. The study was conducted on a "coastal" soil under weather conditions described as "dry," "fair," and "good."

Yield response of corn (bushels acre^{-1}) to applied nitrogen

	Applied N, lb acre^1											
Weather	0	20	40	60	80	100	120	140	160	180	200	220
Dry	14.1	29.8	—	49.3	53.9	54.7	60.2	55.4	59.0	57.6	64.8	66.0
Fair	18.7	43.4	32.0	64.1	69.1	81.0	75.0	85.9	—	83.4	—	—
Good	19.3	40.1	61.7	55.8	82.4	90.8	89.2	86.9	98.6	109.4	90.5	99.7

Data adapted from Baum et al. (1956, table 4.1).

a. Plot yield Y vs. applied N, N, on linear graph paper.
b. Plot the curves on (a) for the model

$$Y = \frac{A}{1 + \exp(1.10 - 0.035N)}$$

where $A = 60$ (dry), 85 (fair), 100 (good) lb acre^{-1}.

c. Discuss the results. Does parameter A appear to account for the effect of weather?

1.8 Consider data for wheat in the table below. The experiments were conducted on sandy soil at Yuma Mesa, AZ.

Yield response of wheat (lb acre^{-1}) to applied nitrogen and irrigation at Yuma Mesa, AZ (1970–71)

Irrigation in.	Applied N lb acre^{-1}	Yield lb acre^{-1}	Irrigation in.	Applied N lb acre^{-1}	Yield lb acre^{-1}
22.4	25	568	41.0	25	519
25.6	100	1176	34.0	100	2276
22.8	175	1152	41.6	175	3147
26.0	250	1818	35.2	250	3654
23.2	325	1614	42.4	325	3988

Data adapted from Hexem and Heady (1978, table 7.7).

a. Plot yield Y vs. applied N, N, on linear graph paper.
b. Plot the curves on (a) for the model.

$$Y = \frac{A}{1 + \exp(2.20 - 0.020N)}$$

where $A = 1800$ lb acre^{-1} (low irrigation) and 4000 lb acre^{-1} (high irrigation).

c. Discuss the results. Does parameter A appear to account for the effect of irrigation?

REFERENCES

Abramowitz, M., and I. A. Stegun. 1965. *Handbook of Mathematical Functions.* Dover Publications, New York.

Bak, P. 1996. *How Nature Works: The Science of Self-Organized Criticality.* Springer-Verlag, New York.

Baum, E. L., E. O. Heady, and J. Blackmore. 1956. *Economic Analysis of Fertilizer Use Data.* Iowa State College Press, Ames, IA.

Born, M. 1956. *Experiment and Theory in Physics.* Dover Publications, New York.

Box, J. F. 1978. *R. A. Fisher: The Life of a Scientist.* John Wiley & Sons, New York.

Bronowski, J. 1973. *The Ascent of Man.* Little, Brown and Co., Boston.

Canine, C. 1995. *Dream Reaper.* Alfred A. Knopf, New York.

Cline, B. L. 1987. *Men Who Made a New Physics*. University of Chicago Press, Chicago.

Feynman, R. 1965. *The Character of Physical Law*. MIT Press, Cambridge, MA.

Ford, A. 1999. *Modeling the Environment*. Island Press, Washington, DC.

Hall, A. D. 1905. *The Book of the Rothamsted Experiments*. John Murray, London.

Hellriegel, H., and H. Wilfarth. 1888. Untersuchungen öber die stickstoffnahrung der gramineen and leguminosen. *Zeitsch. Des Vereins f. d. Rubenzucker-Industrie.*

Hellriegel, H., H. Wilfarth, Roemer, and Wimmer. 1898. Vegetationsversuche öber den kalibedarf einiger pflanzem. *Arb. Deut. Landw. Gesell.*

Hexem, R. W., and E. O. Heady. 1978. *Water Production Functions for Irrigated Agriculture*. Iowa State University Press, Ames, IA.

Holton, G. 1973. *Thematic Origins of Scientific Thought: Kepler to Einstein*. Harvard University Press, Cambridge, MA.

Knowles, F., and J. E. Watkins. 1931. The assimilation and translocation of plant nutrients in wheat during growth. *J. Agric. Sci.* 21:612–637.

Leukel, W. A., J. P. Camp, and J. M. Coleman. 1934. Effect of frequent cutting and nitrate fertilization on the growth behavior and relative composition of pasture grasses. *Florida Agric. Exp. Sta. Bull. 269*. Gainesville, FL.

Lightman, A. 2000. *Great Ideas in Physics*. McGraw-Hill, New York.

Lindley, D. 2001. *Boltzmann's Atom: The Great Debate that Launched a Revolution in Physics*. The Free Press, New York.

McClelland, P. D. 1997. *Sowing Modernity: America's First Agricultural Revolution*. Cornell University Press, Ithaca, NY.

Mitscherlich, E. A. 1909. Das gesets des minimums and das gestez des abnehmenden bodenertages. *Landw. Jahrb.* 38:537–552.

Overman, A. R. 1984. Estimating crop growth rate with land treatment. *J. Env. Engr. Div., Amer. Soc. Civil Engr.* 110:1009–1012.

Overman, A. R. 1998. An expanded growth model for grasses. *Commun. Soil Sci. Plant Anal.* 29:67–85.

Overman, A. R., and D. M. Wilson. 1999. Physiological control of forage grass yield and growth. In: *Crop Yield, Physiology and Processes*, pp. 443–473. D. L. Smith and C. Hamel (eds), Springer-Verlag, Berlin.

Pagels, H. R. 1988. *The Dreams of Reason: The Computer and the Rise of the Sciences of Complexity*. Simon & Schuster, New York.

Penrose, R. 1989. *The Emperor's New Mind: Concerning Computers, Minds, and the Laws of Physics*. Penguin Books, New York.

Polkinghorne, J. 1996. *Beyond Science.* Cambridge University Press, London.

Russell, E. J. 1912. *Soil Conditions and Plant Growth,* 1st ed. Longmans, Green & Co., London.

Russell, E. J. 1937. *Soil Conditions and Plant Growth,* 7th ed. Longmans, Green & Co., London.

Russell, E. J. 1950. *Soil Conditions and Plant Growth,* 8th ed. Longmans, Green & Co., London.

Schlebecker, J. T. 1975. *Whereby We Thrive: A History of American Farming, 1607–1972.* Iowa State University Press, Ames, IA.

van der Paauw, F. 1952. Critical remarks concerning the validity of the Mitscherlich effect law. *Plant and Soil* 4:97–106.

Wick, D. 1995. *The Infamous Boundary: Seven Decades of Heresy in Quantum Physics.* Springer-Verlag, New York.

2

Seasonal Response Models

2.1 BACKGROUND

In Chapter 1 justification was provided for the logistic model over the Mitscherlich model for relating dry matter production to applied nutrients. In this chapter further analysis and application of the logistic model are given. Many details can be found in Overman et al. (1990a, 1990b) and Overman and Wilson (1999). Additional details will be covered in Chapter 4.

2.2 EXTENDED LOGISTIC MODEL

The logistic model can be extended to cover plant nutrient uptake as well as dry matter production. A starting point for this model is three postulates: (1) annual dry matter yield follows logistic response to applied N, (2) annual plant N uptake follows logistic response to applied N, and (3) the N response coefficients are the same for both. Postulate 1 can be written mathematically as

$$Y = \frac{A}{1 + \exp(b - cN)} \qquad [2.1]$$

where Y = seasonal dry matter yield, Mg ha^{-1}; N = applied nitrogen, kg ha^{-1}; A = maximum seasonal dry matter yield, Mg ha^{-1}; b = intercept parameter for dry matter; and c = N response coefficient for dry matter, ha kg^{-1}. Note that the units on parameter c are the inverse of the units on N. Likewise, postulate 2 takes the form

$$N_u = \frac{A_n}{1 + \exp(b_n - c_n N)}$$ [2.2]

where N_u = seasonal plant N uptake, kg ha^{-1}; A_n = maximum seasonal plant N uptake, kg ha^{-1}; b_n = intercept parameter for plant N uptake; c_n = N response coefficient for plant N uptake, ha kg^{-1}.

Now postulates 1 and 2 can be true independently without any obvious connection. Postulate 3 makes a connection with the assumption

$$c_n = c$$ [2.3]

which reduces the number of model parameters from 6 to 5. The importance of this assumption in the plant system will be pointed out shortly. There is some similarity of this postulate to the principle of equivalence of inertial and gravitational masses used by Einstein to develop the theory of general relativity (Hoffmann, 1972, p. 108). Now it follows from Eqs. [2.1] through [2.3] that plant N concentration, $N_c = N_u/Y$, is given by

$$N_c = N_{cm} \left[\frac{1 + \exp(b - cN)}{1 + \exp(b_n - cN)} \right]$$ [2.4]

where $N_{cm} = A_n/A$ = maximum plant N concentration at high applied N. From Eq. [2.4] we obtain the following important mathematical characteristics:

$$N \to +\infty \quad N_c \to N_{cm}$$ [2.5]

$$N = 0 \quad N_c = N_{cm} \left[\frac{1 + \exp(b)}{1 + \exp(b_n)} \right] = N_{c0}$$ [2.6]

$$N \to -\infty \quad N_c \to N_{cm} \exp(b - b_n) = N_{cl}$$ [2.7]

where N_{c0} = plant N concentration at $N = 0$, and N_{cl} = lower plant N concentration at reduced N. For $b_n > b$ we see that

$$N_{cl} < N_{c0} < N_{cm}$$ [2.8]

in ascending order as we would expect for plant N concentration with increasing applied N. This all occurred because of Eq. [2.3], common parameter c. If this were not true, then N_{cl} would either go to zero or blow up. Equations [2.1] and [2.2] can be linearized to the forms

$$\ln\left(\frac{A}{Y} - 1\right) = b - cN \qquad [2.9]$$

$$\ln\left(\frac{A_n}{N_u} - 1\right) = b_n - cN \qquad [2.10]$$

which will be useful for estimating model parameters.

Now we proceed to develop a phase relation between the two dependent variables Y and N_u. This procedure is well established in applied mathematics (Kramer, 1982, p. 356; Williams, 1997, p. 24), as first invented by Maxwell (Ruhla, 1993, p. 101). The first step is to solve Eqs. [2.9] and [2.10] for cN:

$$\ln\left(\frac{A}{Y} - 1\right) - b = -cN = \ln\left(\frac{A_n}{N_u} - 1\right) - b_n \qquad [2.11]$$

Equation [2.11] can be rearranged to the form

$$\ln\left(\frac{A}{Y} - 1\right) = \ln\left(\frac{A_n}{N_u} - 1\right) - \Delta b \qquad [2.12]$$

where

$$\Delta b = b_n - b \qquad [2.13]$$

Equation [2.12] can be rewritten as

$$\frac{A}{Y} - 1 = \left(\frac{A_n}{N_u} - 1\right)\exp(-\Delta b) \qquad [2.14]$$

or

$$Y = \frac{A}{1 + \left(\frac{A_n}{N_u} - 1\right)\exp(-\Delta b)} \qquad [2.15]$$

After a few steps of manipulation Eq. [2.15] takes the form

$$Y = \frac{Y_m N_u}{K_n + N_u} \qquad [2.16]$$

where the new parameters Y_m and K_n are defined by

$$Y_m = \frac{A}{1 - \exp(-\Delta b)} \qquad [2.17]$$

$$K_n = \frac{A_n}{\exp(\Delta b) - 1} \qquad [2.18]$$

According to Eq. [2.16] the relationship between yield Y and plant N uptake N_u should be hyperbolic. A simple rearrangement of Eq. [2.16] gives the relationship of plant N concentration N_c with plant N uptake N_u as

$$N_c = \frac{N_u}{Y} = \frac{K_n}{Y_m} + \frac{N_u}{Y_m} \qquad [2.19]$$

which is linear. Plots of Y and N_c vs. N_u can be used to test the consequences of the extended logistic model directly. Equations [2.17] and [2.18] show the central importance of Δb in determining potential response of the system. According to Eq. [2.16] even if the plant can be induced to take up very high levels of N, dry matter production would still be limited by something else (carbon intake, interception of solar energy, genetics, ?).

2.2.1 Data Analysis

Our next step is to examine a set of data somewhat in detail to illustrate application of the model. We choose the same set of data as used by Overman et al. (1994) originally to present the extended logistic model. Data (Table 2.1) are from the study by Robinson et al. (1988) for dallis-grass (*Paspalum dilatatum* Poir.) grown at Baton Rouge, LA on Olivier silt loam (fine-silty, mixed, thermic Aquic Fragiudalfs). Results are also shown in Fig. 2.1. The first thing to note is that all three variables (Y, N_u, and N_c) increase with increase in applied N, and asymptotically approach maximum values at high N, which is consistent with the model. The second thing to note are the trends in Fig. 2.2, where Y increases with N_u on a somewhat hyperbolic trend, and that N_c vs. N_u appears to follow

Table 2.1 Seasonal dry matter yield, plant N removal, and plant N concentration for dallisgrass grown at Baton Rouge, LA

Applied N (N) kg ha^{-1}	Dry matter yield (Y) Mg ha^{-1}	Plant N removal (N_u) kg ha^{-1}	Plant N concentration (N_c) g kg^{-1}
0	5.33	77	15.7
56	6.56	103	16.2
112	7.97	129	17.0
224	10.53	194	18.9
448	13.21	305	23.4
896	15.34	417	27.5

Data adapted from Robinson et al. (1988).

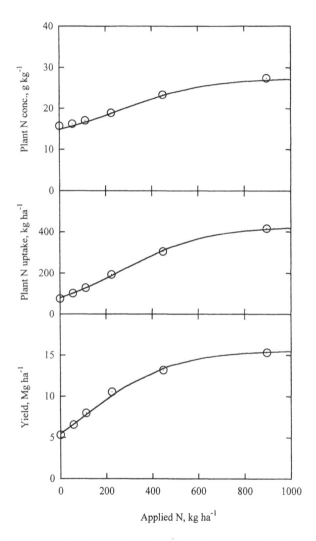

Figure 2.1 Response of dry matter and plant N to applied nitrogen for dallisgrass. Data from Robinson et al. (1988) as discussed by Overman et al. (1994). Curves drawn from Eqs. [2.24] through [2.26].

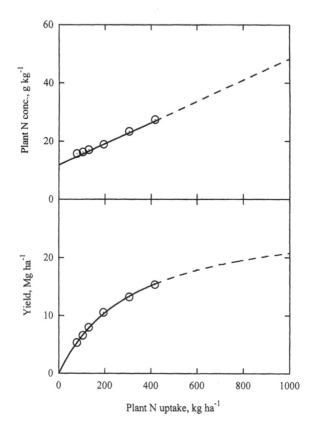

Figure 2.2 Dependence of dry matter and plant N concentration on plant N uptake by dallisgrass for experiment of Robinson et al. (1988). Curve drawn from Eq. [2.27] and line from Eq. [2.28].

a linear relationship. Both effects are predicted by the model. So, we appear to be on the right track. In addition, Eqs. [2.9] and [2.10] suggest an alternative way to plot data, viz. on semilog paper. To do this we need estimates of parameters A and A_n. From Table 2.1 and Fig. 2.1 we select $A = 15.50$ Mg ha^{-1} and $A_n = 425$ kg ha^{-1} by visual inspection. Values of $(A/Y-1)$ and (A_n/N_u-1) are then calculated for each N, which are listed in Table 2.2 and plotted in Fig. 2.3. Results appear to follow straight lines as predicted by Eqs. [2.9] and [2.10]. Furthermore, the lines seem to be approximately parallel as also predicted (common c), with different intercepts (b and b_n). Things look good so far. Linear regression on results from Table 2.2 gives

Table 2.2 Linearized values of yield and plant N uptake for dallisgrass grown at Baton Rouge, LA

Applied N N kg ha^{-1}	$15.50/Y - 1$	$425/N_u - 1$	$15.54/Y - 1$	$430/N_u - 1$
0	1.91	4.52	1.92	4.58
56	1.36	3.13	1.37	3.17
112	0.945	2.29	0.950	2.33
224	0.472	1.19	0.476	1.22
448	0.173	0.393	0.176	0.410
896	0.0104	0.0192	0.0130	0.0312

$$\ln\left(\frac{15.50}{Y} - 1\right) = b - cN = 0.63 - 0.00573N \qquad r = -0.9987 \qquad [2.20]$$

$$\ln\left(\frac{425}{N_u} - 1\right) = b_n - c_n N = 1.53 - 0.00601N \qquad r = -0.9984 \qquad [2.21]$$

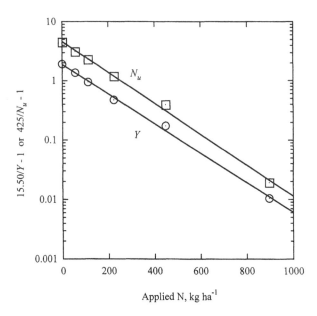

Figure 2.3 Semilog plot of reduced dry matter and plant N uptake for dallisgrass for experiment of Robinson et al. (1988). Lines drawn from Eqs. [2.20] and [2.21].

Several things stand out from Eqs. [2.20] and [2.21]. The correlation coefficients are very high (0.9987 and 0.9984, respectively), the negative values arising from the negative slopes in Fig. 2.3. While the values for c and c_n (0.00573 and 0.00601, respectively) are not identical, we can assume them equal (FAPP) within the accuracy of choosing A and A_n in the first place. Finally, it is apparent that $b_n > b$ (1.53 > 0.63) as expected. At this point we must make a choice, either proceed with parameter values determined from the linearized model and linear regression, repeat the process with new estimates of A and A_n, or perform nonlinear regression on the original model. With new estimates of $A = 15.54$ Mg ha^{-1} and $A_n = 430$ kg ha^{-1}, new linearized ratios in Table 2.2 are obtained, which leads to

$$\ln\left(\frac{15.54}{Y} - 1\right) = b - cN = 0.60 - 0.00549N \qquad r = -0.9990 \qquad [2.22]$$

$$\ln\left(\frac{430}{N_u} - 1\right) = b_n - c_n N = 1.48 - 0.00550N \qquad r = -0.9996 \qquad [2.23]$$

with slightly higher correlation coefficients. Results are shown in Fig. 2.4, where a better fit is apparent.

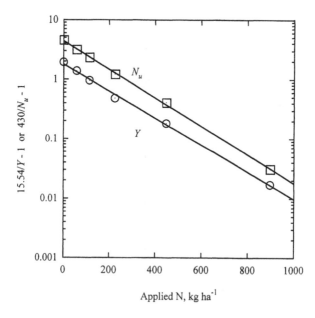

Figure 2.4 Semilog plot of reduced dry matter and plant N uptake for dallisgrass for experiment of Robinson et al. (1988). Lines drawn from Eqs. [2.22] and [2.23].

At this point there are several different ways to proceed. We use analysis of variance (ANOVA) as discussed by Overman et al. (1994) to examine coupling between Eqs. [2.1] and [2.2]. The approach has been outlined by Ratkowsky (1983). Results are listed in Table 2.3. In mode (1) it is assumed that a common A, b, c fits all of the data in Table 2.1 for dry matter yields and plant N uptake, which involves a total of 3 parameters. Since there are 12 data points (6 for yields, and 6 for plant N uptake), this leaves 9 degrees of freedom (df). The residual sum of squares (RSS) between model and data is 8.54 10^{-2}, with a resulting mean sum of squares (MSS) of 9.49 10^{-3}. The best fit of the data by the model will result from fitting yield data and plant N uptake data separately, which is mode (2). In this case we have 6 parameters (3 for yields and 3 for plant N uptake), giving 6 degrees of freedom. Now MSS becomes 2.25 10^{-4}. The difference between modes (1) and (2) gives $RSS^{-2} = 8.41$ 10^{-2} with $9 - 6 = 3$ degrees of freedom for MSS $= 2.80$ 10^{-2}. The variance ratio is given by F $= 2.80$ $10^{-2}/2.25$ $10^{-4} = 124$. The critical value for this case is F(3,6,95) $= 4.76$ from probability tables. It follows that mode (2) gives a significantly better fit of the data than mode (1), which seems almost self-evident since the magnitude of numbers for yields and plant N uptake are so different. The next obvious step is to try mode (3) with different values of A and A_n, but with common b and c. Now we obtain F $= 100$, which is slightly better than mode (1) but still greatly exceeds the critical value of F(2,6,95) $= 5.14$. Finally, we assume mode (4) with individual A and b, but with common c. This gives F $= 1.6$ compared to the critical value of F(1,6,95) $= 5.99$. It follows that mode (4)

Table 2.3 Analysis of variance for model parameters for dallisgrass grown at Baton Rouge, LA

Mode		Parameters estimated	df	Residual sum of squares	Mean sum of squares	F
(1)	Common A, b, c	3	9	8.54 10^{-2}	9.49 10^{-3}	—
(2)	Individual A, b, c	6	6	1.35 10^{-3}	2.25 10^{-4}	—
(1)–(2)		—	3	8.41 10^{-2}	2.80 10^{-2}	124
(3)	Individual A Common b, c	4	8	4.49 10^{-2}	5.61 10^{-3}	—
(3)–(2)		—	2	4.48 10^{-2}	2.24 10^{-2}	100
(4)	Individual A, b Common c	5	7	1.72 10^{-3}	2.46 10^{-4}	—
(4)–(2)		—	1	3.70 10^{-4}	3.70 10^{-4}	1.6

F(3,6,95) $= 4.76$; F(2,6,95) $= 5.14$; F(1,6,95) $= 5.99$.

and mode (2) are not significantly different, supporting the common *c*. (*Note*: There are mistakes in table 1 of Overman et al., 1994).

Without going into the details of the nonlinear regression procedure (which are covered in Chapter 6), let us simply use values from Overman et al. (1994) listed in Table 2.4 for mode (4) above, which are not very different from those estimated by linear regression. The curves in Fig. 2.1 are drawn from the equations

$$Y = \frac{15.60}{1 + \exp(0.61 - 0.0054N)} \qquad [2.24]$$

$$N_u = \frac{427}{1 + \exp(1.45 - 0.0054N)} \qquad [2.25]$$

$$N_c = 27.5 \left[\frac{1 + \exp(0.61 - 0.0054N)}{1 + \exp(1.45 - 0.0054N)} \right] \qquad [2.26]$$

and the curve and line, respectively, in Fig. 2.2 from

$$Y = \frac{27.5 N_u}{325 + N_u} \qquad [2.27]$$

$$N_c = 11.8 + 0.0364 N_u \qquad [2.28]$$

It can be seen that the model describes the results quite well.

A number of other points should be made about these results. The maximum incremental yield response (slope), which provides one half of maximum yield to applied N, occurs at $N_{1/2} = b/c = 0.61/0.0054 = 110$ kg ha^{-1}, which gives $Y = 7.8$ Mg ha^{-1} and $N_u = 127$ kg ha^{-1}, for an N recovery of 115%. Similarly, for maximum incremental plant N uptake we obtain $(N_{1/2})_n = b_n/c = 1.45/0.0054 = 270$ kg ha^{-1}, which gives $Y = 10.9$ Mg ha^{-1} and $N_u = 214$ kg ha^{-1}, for an N recovery of 80%. Since $\Delta b = b_n - b = 1.45 - 0.61 = 0.84$, it follows from Eq. [2.17] that the

Table 2.4 Model parameters determined by nonlinear regression for dallisgrass grown at Baton Rouge, LA

Parameter	A Mg ha^{-1}	A_n kg ha^{-1}	b	b_n	c ha kg^{-1}	N_{cm} g kg^{-1}	Y_m Mg ha^{-1}	K_n kg ha^{-1}
Value	15.60	427	0.61	1.45	0.0054	27.5	27.5	325

Parameter values from Overman et al. (1994).

ratio of potential maximum yield, Y_m, to maximum yield, A, is $Y_m/A = 1.75$. This suggests that something other than plant N uptake ultimately becomes limiting. Analysis by Overman (1995b) of data from Allen et al. (1990) showed that the present atmospheric CO_2 level produced approximately 50% of potential leaf area for soybeans [*Glycine max* (L.) Merr.]. Perhaps carbon intake from the atmosphere is one limiting factor on dry matter production for forage grass. This issue is discussed further in Section 2.6.

2.2.2 Discussion

Now we can discuss strengths and criticisms of the model. First, the strengths: The model is relatively easy to use, easily run on a pocket calculator. In fact it is no more difficult than a polynomial (say quadratic). It is mathematically well behaved, meaning that it is bounded between zero and a maximum, is monotone increasing ($dY/dN > 0$) over the range of N, and gives the correct behavior for N_c vs. N. It predicts the correct behavior for Y and N_c vs. N_u, which is very significant. So, what are the criticisms of the model? Mainly that it is not "mechanistic." That is certainly true. However, there is a rational basis to the model (Overman, 1995a). It is not, as some have stated, "just curve fitting." That statement is without foundation. Furthermore, there is little or no hope of developing a truly mechanistic model of crop growth for two reasons: (1) The geometry of roots and tops is simply too complex to describe, even for one plant, and (2) the molecular processes would at best require statistical and quantum mechanics to describe the fundamental processes at work (Johnson et al., 1974), both of which are hopelessly complex for a whole plant or field of plants. Let us recall that we can't even solve the Navier-Stokes equation of fluid flow through soil due to complex geometry of the flow paths and the liquid/solid boundary; we use the empirical Darcy equation instead. However, some progress has been made in coupling applied nutrients, available soil nutrients, accumulation in roots and tops, and linkage to CO_2 in the canopy (Overman, 1995b). We believe that there is some confusion between "compartmental" and "mechanistic" models. These are not the same thing. In our opinion we should keep an open mind on approaches to crop models, without trying to cut off promising alternatives. Perhaps in the present climate this hope is a little naïve! There is the longer range view to scholarship, which we shall try to follow. The role of intuition in developing models should not be underestimated (Hoffman, 1972, p. 193).

The term "N concentration" is a somewhat mistaken definition, since we do not measure concentration in the conventional chemical sense. Instead, mass of plant N is determined for a specified mass of dry matter, as defined in Eq. [2.19]. Perhaps the term "specific N" would be more appropriate and consistent with other branches of science. As often happens, terminology becomes entrenched and is difficult to change.

It could be argued that a set of data was chosen which just happened to follow logistic behavior; in other words, this is circumstantial. In fact, a large number of data sets have been analyzed and shown to follow the pattern described here. Ultimately, we should remember that a model can only be proven wrong, and never "true" (Stewart, 1989, p. 174). The evidence overwhelmingly supports use of the logistic model. It has also been extended to cover applied P and K as well as applied N (Overman and Wilkinson, 1995). Controversy over which model to use is likely to go on indefinitely. Individuals and agencies have a lot at stake in this subject. Of course, this nothing new in science and engineering. A lot of determination is called for in such circumstances.

2.2.3 Alternative Model Development

As an alternative to the procedure described above, the model can be developed from two postulates. Postulate 1: Plant N uptake follows logistic response to applied N, which can be stated mathematically as

$$N_u = \frac{A_n}{1 + \exp(b_n - c_n N)} \tag{2.29}$$

where N_u = seasonal plant N uptake, kg ha^{-1}; A_n = maximum seasonal plant N uptake, kg ha^{-1}; b_n = intercept parameter for plant N uptake; c_n = N response coefficient for plant N uptake, ha kg^{-1}. Postulate 2: Dry matter yield follows hyperbolic dependence on plant N uptake, which takes the form

$$Y = \frac{Y_m N_u}{K_n + N_u} \tag{2.30}$$

where Y = seasonal dry matter yield, Mg ha^{-1}; Y_m = potential maximum yield, Mg ha^{-1}; and K_n = system response coefficient, kg ha^{-1}.

Several consequences follow from Eqs. [2.29] and [2.30]. Perhaps the simplest of these is for plant N concentration N_c obtained by rearrangement of Eq. [2.30] to give

$$N_c = \frac{N_u}{Y} = \frac{K_n}{Y_m} + \frac{N_u}{Y_m} \tag{2.31}$$

which shows linear dependence of plant N concentration on plant N uptake. Other consequences now follow. Substitution of Eq. [2.29] into Eq. [2.30] leads to

$$Y = \frac{Y_m \frac{A_n}{1+\exp(b_n-c_nN)}}{K_n + \frac{A_n}{1+\exp(b_n-c_nN)}} = \frac{Y_m A_n}{A_n + K_n + K_n \exp(b_n - c_nN)}$$

$$= \frac{Y_m \frac{A_n}{A_n+K_n}}{1 + \left[\frac{K_n}{A_n+K_n}\right] \exp(b_n - c_nN)} \qquad [2.32]$$

or

$$Y = \frac{A}{1 + \exp(b - c_nN)} \qquad [2.33]$$

where

$$A = Y_m \frac{A_n}{A_n + K_n} \qquad [2.34]$$

and

$$\exp(b) = \frac{K_n}{K_n + A_n} \exp(b_n) \qquad [2.35]$$

or

$$b = b_n - \ln\left(1 + \frac{A_n}{K_n}\right) = \text{intercept parameter for dry matter} \qquad [2.36]$$

It is convenient to define a new parameter

$$\Delta b = b_n - b = \ln\left(1 + \frac{A_n}{K_n}\right) \qquad [2.37]$$

Since $A_n/K_n \geq 0$, then $\Delta b \geq 0, b_n \geq b$. Equations [2.34] and [2.36] can be rearranged to give

$$Y_m = \frac{A}{1 - \exp(-\Delta b)} \qquad [2.38]$$

$$K_n = \frac{A_n}{\exp(\Delta b) - 1} \qquad [2.39]$$

According to Eq. [2.33] dry matter yield follows logistic response to applied N. This means that even if only yield vs. applied N is reported, it is still possible to use the logistic model for analysis without plant N uptake data. Since plant N concentration is defined as $N_c = N_u/Y$, then plant N concentration is given by the ratio of logistic equations

$$N_c = N_{cm}\left[\frac{1 + \exp(b - c_nN)}{1 + \exp(b_n - c_nN)}\right] \qquad [2.40]$$

where

$$N_{cm} = \frac{A_n}{A} = \text{maximum plant N concentration, g kg}^{-1} \qquad [2.41]$$

Equations [2.29] and [2.33] can be written in the linearized forms

$$\ln\left(\frac{A_n}{N_u} - 1\right) = b_n - c_n N \qquad [2.42]$$

$$\ln\left(\frac{A}{Y} - 1\right) = b - c_n N \qquad [2.43]$$

According to Eqs. [2.42] and [2.43], a semilog plot of $(A_n/N_u - 1)$ and $(A/Y - 1)$ vs. N should give two parallel lines with slopes c_n and intercepts b_n and b, respectively.

This alternative approach eliminates the question of two c values (there is only one) and clarifies the difference in the two b values. It is also founded on only two postulates instead of three, and is therefore more compact in statement, which also invokes "Occam's razor" (Asimov, 1966, p. 5, 235; Kline, 1981, p. 403).

Equation [2.30] can be rearranged into the double reciprocal form

$$\frac{1}{Y} = \frac{1}{Y_m} + \frac{K_n}{Y_m} \frac{1}{N_u} \qquad [2.44]$$

Data from Table 2.1 are given in this form in Table 2.5 and shown in Fig. 2.5. Linear regression of $1/Y$ vs. $1/N_u$ gives

$$\frac{1}{Y} = 0.0365 + 11.66 \, \frac{1}{N_u} \qquad r = 0.9994 \qquad [2.45]$$

which leads to $Y_m = 27.4$ Mg ha^{-1} and $K_n = 319$ kg ha^{-1}. These values compare very closely with $Y_m = 27.5$ Mg ha^{-1} and $K_n = 325$ kg ha^{-1} from Table 2.4 for nonlinear regression of two logistic equations with

Table 2.5 Reciprocal values of yield and plant N uptake for dallisgrass grown at Baton Rouge, LA

Applied N kg ha^{-1}	$1/N_u$ ha kg^{-1}	$1/Y$ ha Mg^{-1}
0	0.0130	0.188
56	0.00971	0.152
112	0.00775	0.125
224	0.00515	0.0950
448	0.00328	0.0757
896	0.00240	0.0652

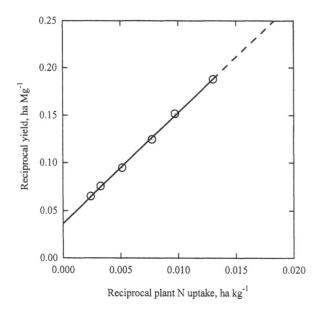

Figure 2.5 Double reciprocal plot of dry matter dependence on plant N uptake for dallisgrass for experiment of Robinson et al. (1988). Line drawn from Eq. [2.45].

common c. One could also perform nonlinear regression on Eq. [2.30] to obtain estimates of Y_m and K_n.

Equation [2.33] can also be written in the form

$$Y = \frac{A}{1 + \exp\left(\frac{N_{1/2} - N}{N'}\right)} \qquad [2.46]$$

where $N_{1/2} = b/c_n =$ applied N for $Y = A/2$, kg ha^{-1}; and $N' = 1/c_n =$ characteristic N, kg ha^{-1}. One advantage of this form is that the parameters $N_{1/2}$ and N have units of N in kg ha^{-1}. It is common practice in science and engineering to define characteristic quantities in systems which follow an exponential relationship (Silverman, 1998, p. 22).

2.2.4 Application to Vegetable Crops

The extended logistic model can also be applied to vegetable crops. Response of sweet corn [*Zea mays*] to applications of organic fertilizer (mushroom compost) has been reported by Rhoads and Olson (1995). Results are given in Table 2.6 and shown in Fig. 2.6, where the curves are drawn from

Table 2.6 Seasonal fresh weight yield, plant N removal, and plant N concentration for sweet corn grown at Quincy, FL

N kg ha^{-1}	Y Mg ha^{-1}	N_u kg ha^{-1}	N_c g kg^{-1}	$11.9/Y - 1$	$158/N_u - 1$
0	3.7	35	9.4	2.22	3.51
322	7.6	81	10.6	0.566	0.951
644	11.4	147	12.9	0.0439	0.0748
1288	11.8	156	13.2	0.00848	0.0128

Data for fresh weights and plant N adapted from Rhoads and Olson (1995).

$$Y = \frac{11.9}{1 + \exp(0.58 - 0.00445N)} \tag{2.47}$$

$$N_u = \frac{158}{1 + \exp(1.08 - 0.00445N)} \tag{2.48}$$

$$N_c = \left[\frac{1 + \exp(0.58 - 0.00445N)}{1 + \exp(1.08 - 0.00445N)} \right] \tag{2.49}$$

The linearized forms of Eqs. [2.47] and [2.48] given by

$$\ln\left(\frac{11.9}{Y} - 1\right) = b - c_n N = 0.58 - 0.00445N \qquad r = -0.974 \tag{2.50}$$

$$\ln\left(\frac{158}{N_u} - 1\right) = b_n - c_n N = 1.08 - 0.00445N \qquad r = -0.977 \tag{2.51}$$

are used to draw the lines in Fig. 2.7. Equations [2.17] and [2.18] are used to calculate the phase equation

$$Y = \frac{Y_m N_u}{K' + N_u} = \frac{30.2 N_u}{244 + N_u} \tag{2.52}$$

shown in Fig. 2.8, where the line for plant N concentration is drawn from

$$N_c = \frac{N_u}{Y} = \frac{K_n}{Y_m} + \frac{1}{Y_m} N_u = 8.1 + 0.0331 N_u \tag{2.53}$$

This study also included tomato (*Lycopersicon esculentum* Mill) and squash [*Cucurbita pepo* L.). Results are given in Table 2.7 and shown Fig. 2.9, where the yield response curves are drawn from

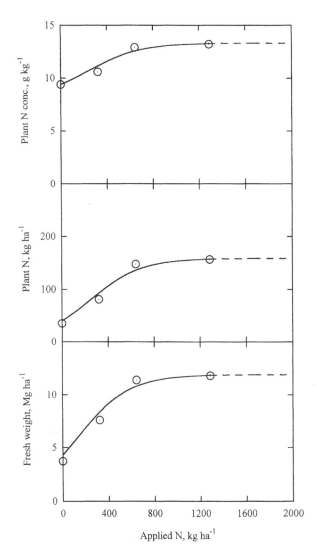

Figure 2.6 Response of fresh weight and plant N to applied N for sweet corn for experiment of Rhoads and Olson (1995). Curves drawn from Eqs. [2.47] through [2.49].

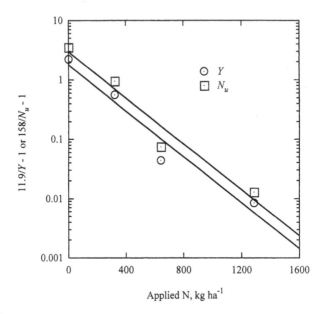

Figure 2.7 Semilog plot of sweet corn response to applied N for experiment of Rhoads and Olson (1995). Lines drawn from Eqs. [2.50] and [2.51] for fresh weight and plant N uptake, respectively.

Tomato: $Y = \dfrac{55.9}{1 + \exp(0.49 - 0.00445N)}$ [2.54]

Squash: $Y = \dfrac{10.3}{1 + \exp(2.20 - 0.00445N)}$ [2.55]

where parameter $c_n = 0.00445$ ha kg^{-1} is used for all three crops on the same soil (unspecified). The model provides excellent description of response of vegetables to applied N.

The logistic model has also been applied to response of lettuce (*Lactuca sativa*) to applied N (Willcutts et al., 1998).

2.3   EXTENDED MULTIPLE LOGISTIC MODEL

In the previous section we established the applicability of the extended logistic model for dry matter and plant N uptake. Overman et al. (1995) have expanded its application to the elements P and K individually. Now if the model applies for N, P, and K separately, then the question logically arises as to the proper form for combinations of N, P, and K. In experimental work

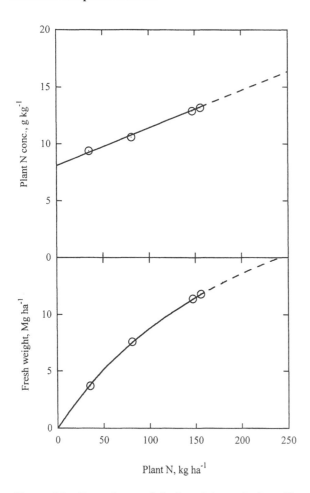

Figure 2.8 Dependence of fresh weight and plant N concentration on plant N uptake by sweet corn for experiment of Rhoads and Olson (1995). Curve drawn from Eq. [2.52] and line from Eq. [2.53].

such designs are called factorial. If dependence of dry matter production on each element is independent of the other two elements, then one can argue on the grounds of joint probability (Ruhla, 1993) that the multiple logistic model should involve the product of each term and take the form

$$Y = \frac{A}{[1 + \exp(b_n - c_n N)][1 + \exp(b_p - c_p P)][1 + \exp(b_k - c_k K)]} \qquad [2.56]$$

Table 2.7 Seasonal fresh weight response to applied N for tomato and squash at Quincy, FL

N kg ha^{-1}	Y, Mg ha^{-1} Tomato	Y, Mg ha^{-1} Squash	55.9/Y − 1 Tomato	10.3/Y − 1 Squash
0	23.4	1.0	1.39	9.30
322	37.8	3.6	0.479	1.86
644	50.8	6.1	0.100	0.689
1288	55.6	10.0	0.00540	0.0300

Data for fresh weights from Rhoads and Olson (1995).

where Y = dry matter yield, Mg ha^{-1}; N, P, K = applied N, P, K, respectively, kg ha^{-1}; A = maximum dry matter yield at high N, P, K, Mg ha^{-1}; b_n, b_p, b_k = intercept parameters of dry matter for N, P, K, respectively; c_n , c_p, c_k = response coefficient for N, P, K, respectively, ha kg^{-1}. Note that for fixed levels of two of the elements, Eq. [2.56] reduces to the simple logistic equation in which the effective A depends upon the levels of the two fixed elements. Overman et al. (1991) have applied the multiple logistic model for response of coastal bermudagrass (*Cynodon dactylon* L.) to applied N, P, and K at Watkinsville, GA (Carreker et al., 1977) on Cecil sandy loam (clayey, kaolinitic, thermic Typic Hapludult). The experimental design was a complete factorial of NxPxK = 4 × 4 × 4 = 64 combinations. Resulting model parameters were: $A = 15.54$ Mg ha^{-1}; $b_n = 0.462$; $b_p = -1.58$; $b_k = -1.58$; $c_n = 0.0122$ ha kg^{-1}; $c_p = 0.0410$ ha kg^{-1}; $c_k = 0.0212$ ha kg^{-1}.

By analogy with Eq. [2.56] we assume that dependence of plant nutrient uptake upon applied N, P, and K is also described by the triple logistic equation

$$N_u = \frac{A'}{[1 + \exp(b'_n - c_n N)][1 + \exp(b'_p - c_p P)][1 + \exp(b'_k - c_k K)]} \quad [2.57]$$

where N_u = plant nutrient uptake for N, P, or K, respectively, kg ha^{-1}; N, P, K = applied N, P, K, respectively, kg ha^{-1}; A = maximum plant nutrient uptake for N, P, or K, respectively at high N, P, K, kg ha^{-1}; b'_n, b'_p, b'_k = intercept parameters of plant nutrient uptake for N, P, K, respectively; c_n, c_p, c_k = response coefficient for N, P, K, respectively, ha kg^{-1}. Overman and Wilkinson (1995) analyzed data (Adams et al., 1966) from the same factorial experiment for coastal bermudagrass described above. Examination of the data for plant nutrient concentrations led to the conclusion that, to first

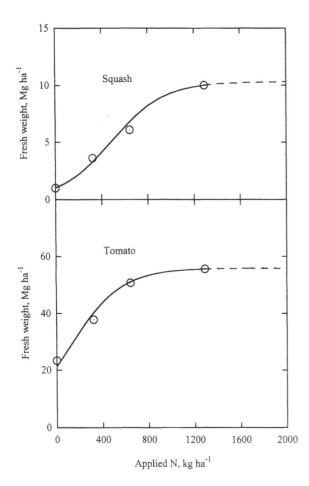

Figure 2.9 Response of fresh weight and plant N to applied N for tomato and squash for experiment of Rhoads and Olson (1995). Curves drawn from Eqs. [2.54] and [2.55].

approximation at least, it was reasonable to assume that concentration of a given element was independent of the applied levels of the other two elements. From this assumption, it follows that the concentration equations are given by

$$N_c = N_{cm} \left[\frac{1 + \exp(b_n - c_n N)}{1 + \exp(b'_n - c_n N)} \right] \qquad [2.58]$$

$$P_c = P_{cm}\left[\frac{1 + \exp(b_p - c_p P)}{1 + \exp(b'_p - c_p P)}\right] \qquad [2.59]$$

$$K_c = K_{cm}\left[\frac{1 + \exp(b_k - c_k K)}{1 + \exp(b'_k - c_k K)}\right] \qquad [2.60]$$

where N_c = plant N concentration, g kg^{-1}; P_c = plant P concentration, g kg^{-1}; K_c = plant K concentration, g kg^{-1}; N_{cm} = maximum plant N concentration, g kg^{-1}; P_{cm} = maximum plant P concentration, g kg^{-1}; K_{cm} = maximum plant K concentration, g kg^{-1}. A summary of the model parameters is given in Table 2.8. In this case maximum plant nutrient concentrations are : N_{cm} = 24.5 g kg^{-1}; P_{cm} = 2.4 g kg^{-1}; K_{cm} = 17.4 g kg^{-1}.

Next we examine data from a study in which levels of N, P, and K were all varied together and in constant ratio. This case presents a special challenge in determining the fraction of response due to each element. Data from a field study by Adams et al. (1967) for coastal and common bermudagrass grown at Watkinsville, GA on Cecil sandy loam are listed in Table 2.9. Fortunately, measurements of both dry matter and plant nutrient uptake were made. We make the simplifying assumptions that the b and c parameters are the same for the two grasses, and that all the difference occurs in the A and A' parameters. Furthermore, estimates of the b and c parameters are taken from Table 2.8 for the study of coastal bermudagrass. The first step is to standardize yield and plant N uptake over applied P and K from

$$Y^* = Y[1 + \exp(b_p - c_p P)][1 + \exp(b_k - c_k K)] = \frac{A}{1 + \exp(b_n - c_n N)} \qquad [2.61]$$

$$N_u^* = N_u[1 + \exp(b_p - c_p P)][1 + \exp(b_k - c_k K)] = \frac{A'}{1 + \exp(b'_n - c_n N)} \qquad [2.62]$$

Table 2.8 Model parameters for coastal bermudagrass grown at Watkinsville, GA

Parameter	Dry matter	N removal	P removal	K removal
A, kg ha^{-1}	15.540	380	37	270
b_n	0.462	1.25	0.462	0.462
b_p	−1.58	−1.58	−0.65	−1.58
b_k	−1.58	−1.58	−1.58	0.30
c_n, ha kg^{-1}	0.0122	0.0122	0.0122	0.0122
c_p, ha kg^{-1}	0.0410	0.0410	0.0410	0.0410
c_k, ha kg^{-1}	0.0212	0.0212	0.0212	0.0212

Parameters from Overman and Wilkinson (1995).

Table 2.9 Common and coastal bermudagrass response to applied nutrients grown at Watkinsville, GA

Grass	N kg ha^{-1}	P kg ha^{-1}	K kg ha^{-1}	Y Mg ha^{-1}	N_u kg ha^{-1}	P_u kg ha^{-1}	K_u g kg^{-1}	N_c g kg^{-1}	P_c g kg^{-1}	K_c g kg^{-1}
Common	0	0	0	2.33	39	5	31	16.7	2.1	13.3
bermuda	112	24	46	5.31	113	12	87	21.3	2.3	16.4
	224	48	92	8.27	201	21	148	24.3	2.5	17.9
	448	96	184	10.33	296	30	228	28.7	2.9	22.1
Coastal	0	0	0	3.74	58	8	48	15.5	2.1	12.8
bermuda	112	24	46	8.92	151	19	136	16.9	2.1	15.2
	224	48	92	12.26	240	29	200	19.6	2.4	16.3
	448	96	184	16.03	400	41	337	25.0	2.6	21.0

Data adapted from Adams et al. (1967).

so that Y^* and N_u^* become functions of applied N only. Results obtained from Eqs. [2.61] and [2.62] are listed in Table 2.10 and shown in Fig. 2.10. The curves in Fig. 2.10 are drawn from

$$Y^* = \frac{A}{1 + \exp(0.80 - 0.0100N)} \qquad [2.63]$$

$$N_u^* = \frac{A'}{1 + \exp(1.50 - 0.0100N)} \qquad [2.64]$$

$$N_c^* = N_{cm} \left[\frac{1 + \exp(0.80 - 0.0100N)}{1 + \exp(1.50 - 0.0100N)} \right] \qquad [2.65]$$

Table 2.10 Standardized yield and plant N uptake response to applied N for common and coastal bermudagrass grown at Watkinsville, GA

Grass	N kg ha^{-1}	Y Mg ha^{-1}	Y^* Mg ha^{-1}	N_u kg ha^{-1}	N_u^* kg ha^{-1}	N_c^* g kg^{-1}
Common	0	2.33	3.39	39	57	16.7
bermuda	112	5.31	6.16	113	131	21.3
	224	8.27	8.76	201	213	24.3
	448	10.33	10.42	296	298	28.7
Coastal	0	3.74	5.44	58	84	15.5
bermuda	112	8.92	10.35	151	175	16.9
	224	12.26	12.98	240	254	19.6
	448	16.03	16.16	400	403	25.0

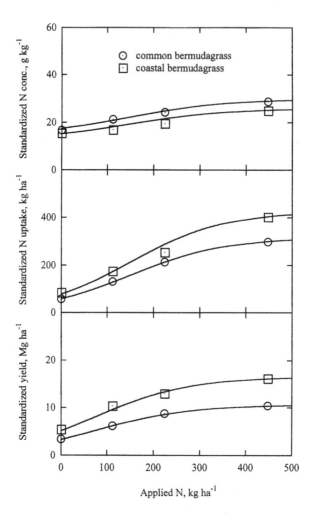

Figure 2.10 Dependence of standardized yield and plant N uptake on applied N for coastal and common bermudagrass. Data from Adams et al. (1967). Curves drawn from Eqs. [2.63] through [2.65].

where $A = 10.60$ Mg ha^{-1}, $A' = 315$ kg ha^{-1}, and $N_{cm} = 29.7$ g kg^{-1} for common bermudagrass and $A = 16.50$ Mg ha^{-1}, $A' = 425$ kg ha^{-1}, and $N_{cm} = 25.8$ g kg^{-1} for coastal bermudagrass. The phase plot for N is shown in Fig. 2.11, where the curves are drawn from

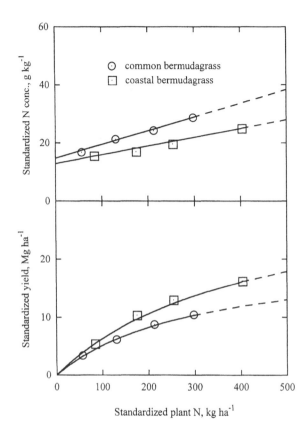

Figure 2.11 Dependence of standardized dry matter on standardized plant N uptake for coastal and common bermudagrass for experiment of Adams et al. (1967). Curves drawn from Eqs. [2.66] and [2.67]; lines drawn from Eqs. [2.68] and [2.69].

Common bermudagrass: $\quad Y^* = \dfrac{21.1\,N_u^*}{311 + N_u^*}$ [2.66]

Coastal bermudagrass: $\quad Y^* = \dfrac{32.8\,N_u^*}{419 + N_u^*}$ [2.67]

and the lines are drawn from

Common bermudagrass: $\quad N_c^* = 14.7 + 0.0474\,N_u^*$ [2.68]

Coastal bermudagrass: $\quad N_c^* = 12.8 + 0.0305\,N_u^*$ [2.69]

Results in Figs. 2.10 and 2.11 support the model very well.

A similar procedure is now used to standardize over applied N and K, which leads to

$$Y^* = Y[1 + \exp(b_n - c_n N)][1 + \exp(b_k - c_k K)] = \frac{A}{1 + \exp(b_p - c_p P)} \quad [2.70]$$

$$P_u^* = P_u[1 + \exp(b_n - c_n N)][1 + \exp(b_k - c_k K)] = \frac{A'}{1 + \exp(b_p' - c_p P)} \quad [2.71]$$

so that Y^* and P_u^* become functions only of applied P. Results obtained from Eqs. [2.70] and [2.71] are listed in Table 2.11 and shown in Fig. 2.12. The curves in Fig. 2.12 are drawn from

$$Y^* = \frac{A}{1 + \exp(-1.58 - 0.0410P)} \quad [2.72]$$

$$P_u^* = \frac{A'}{1 + \exp(-0.65 - 0.0410P)} \quad [2.73]$$

$$P_c^* = P_{cm}\left[\frac{1 + \exp(-1.58 - 0.0410P)}{1 + \exp(-0.65 - 0.0410P)}\right] \quad [2.74]$$

where $A = 10.60$ Mg ha^{-1}, $A' = 30$ kg ha^{-1}, and $P_{cm} = 2.8$ g kg^{-1} for common bermudagrass and $A = 16.50$ Mg ha^{-1}, $A' = 43$ kg ha^{-1}, and $P_{cm} = 2.6$ g kg^{-1} for coastal bermudagrass. The phase plot for P is shown in Fig. 2.13, where the curves are drawn from

Table 2.11 Standardized yield and plant P uptake response to applied P for common and coastal bermudagrass grown at Watkinsville, GA

Grass	P kg ha^{-1}	Y Mg ha^{-1}	Y^* Mg ha^{-1}	P_u kg ha^{-1}	P_u^* kg ha^{-1}	P_c^* g kg^{-1}
Common	0	2.33	9.06	5	19	2.1
bermuda	24	5.31	9.88	12	22	2.3
	48	8.27	10.53	21	27	2.5
	196	10.33	10.63	30	31	2.9
Coastal	0	3.74	14.55	8	31	2.1
bermuda	24	8.92	16.59	19	35	2.1
	48	12.26	15.61	29	37	2.4
	196	16.03	16.50	41	42	2.6

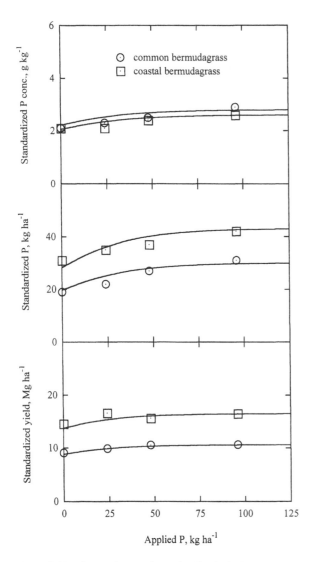

Figure 2.12 Dependence of standardized yield and plant P uptake on applied P for coastal and common bermudagrass. Data from Adams et al. (1967). Curves drawn from Eqs. [2.72] through [2.74].

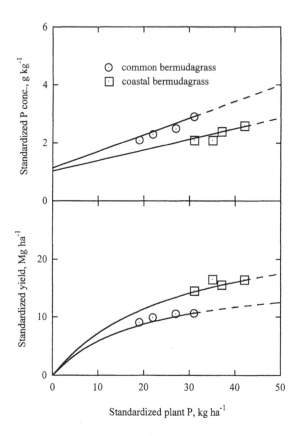

Figure 2.13 Dependence of standardized dry matter on standardized plant P uptake for coastal and common bermudagrass for experiment of Adams et al. (1967). Curves drawn from Eqs. [2.75] and [2.76]; lines drawn from Eqs. [2.77] and [2.78].

Common bermudagrass: $Y^* = \dfrac{17.5\, P_u^*}{20 + P_u^*}$ [2.75]

Coastal bermudagrass: $Y^* = \dfrac{27.3\, P_u*}{28 + P_u^*}$ [2.76]

and the lines are drawn from

Common bermudagrass: $P_c^* = 1.14 + 0.0571\, P_u^*$ [2.77]

Coastal bermudagrass: $P_c^* = 1.03 + 0.0366\, P_u^*$ [2.78]

Results in Figs. 2.12 and 2.13 support the model very well.

Finally, a similar procedure is used to standardize over applied N and P, which leads to

$$Y^* = Y[1 + \exp(b_n - c_n N)][1 + \exp(b_p - c_p P)] = \frac{A}{1 + \exp(b_k - c_k K)} \quad [2.79]$$

$$K_u^* = K_u[1 + \exp(b_n - c_n N)][1 + \exp(b_p - c_p P)] = \frac{A'}{1 + \exp(b_k' - c_k K)} \quad [2.80]$$

so that Y^* and K_u^* become functions only of applied K. Results obtained from Eqs. [2.79] and [2.80] are listed in Table 2.12 and shown in Fig. 2.14. The curves in Fig. 2.14 are drawn from

$$Y^* = \frac{A}{1 + \exp(-1.58 - 0.0212K)} \quad [2.81]$$

$$K_u^* = \frac{A'}{1 + \exp(0.30 - 0.0212K)} \quad [2.82]$$

$$K_c^* = K_{cm}\left[\frac{1 + \exp(-1.58 - 0.0212K)}{1 + \exp(0.30 - 0.0212K)}\right] \quad [2.83]$$

where $A = 10.60$ Mg ha^{-1}, $A' = 250$ kg ha^{-1}, and $K_{cm} = 23.6$ g kg^{-1} for common bermudagrass and $A = 16.50$ Mg ha^{-1}, $A' = 350$ kg ha^{-1}, and $K_{cm} = 21.2$ g kg^{-1} for coastal bermudagrass. The phase plot for K is shown in Fig. 2.15, where the curves are drawn from

$$\text{Common bermudagrass:} \quad Y^* = \frac{12.5\, K_u*}{45 + K_u*} \quad [2.84]$$

$$\text{Coastal bermudagrass:} \quad Y^* = \frac{19.5\, K_u^*}{63 + K_u^*} \quad [2.85]$$

Table 2.12 Standardized yield and plant K uptake response to applied K for common and coastal bermudagrass grown at Watkinsville, GA

Grass	K kg ha^{-1}	Y Mg ha^{-1}	Y^* Mg ha^{-1}	K_u kg ha^{-1}	K_u^* kg ha^{-1}	K_c^* g kg^{-1}
Common	0	2.33	9.06	31	121	13.3
bermuda	46	5.31	9.87	87	162	16.4
	92	8.27	10.52	148	188	17.9
	184	10.33	10.63	228	235	22.1
Coastal	0	3.74	14.55	48	187	12.8
bermuda	46	8.92	16.58	136	253	15.2
	92	12.26	15.60	200	255	16.3
	184	16.03	16.50	337	347	21.0

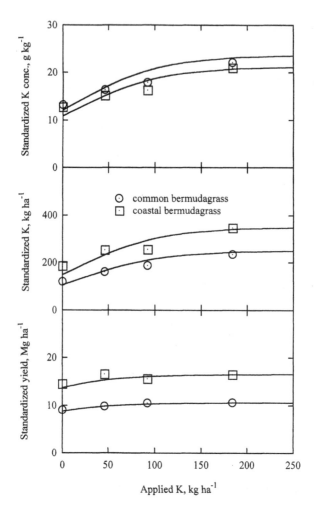

Figure 2.14 Dependence of standardized yield and plant K uptake on applied K for coastal and common bermudagrass. Data from Adams et al. (1967). Curves drawn from Eqs. [2.81] through [2.83].

and the lines are drawn from

Common bermudagrass: $K_c^* = 3.60 + 0.0800 \, K_u^*$ [2.86]

Coastal bermudagrass: $K_c^* = 3.23 + 0.0513 \, K_u^*$ [2.87]

Results in Figs. 2.14 and 2.15 support the model very well. A summary of model parameters for common and coastal bermudagrass is given in

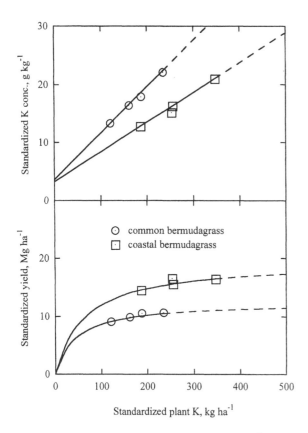

Figure 2.15 Dependence of standarized dry matter on standardized plant K uptake for coastal and common bermudagrass for experiment of Adams et al. (1967). Curves drawn from Eqs. [2.84] and [2.85]; lines drawn from Eqs. [2.86] and [2.87].

Table 2.13. Estimates of system response by the model are given in Table 2.14.

It is now apparent that most of the response of dry matter production in the system was due to applied N, and relatively little to applied P and K. If the coefficients for the multiple logistic model had not been available from previous work, then an approximation would be to attribute all of the response to applied N. Results of this approach are shown in Fig. 2.16, where the curves are drawn from

$$Y = \frac{A}{1 + \exp(1.20 - 0.0100N)} \qquad [2.88]$$

Table 2.13 Model parameters for common and coastal bermudagrass grown at Watkinsville, GA

Grass	Element	A Mg ha^{-1}	A' kg ha^{-1}	b	b'	Y_m Mg ha^{-1}	K' kg ha^{-1}
Common	N	10.60	315	0.80	1.50	21.1	311
bermuda	P		30	−1.58	−0.65	17.5	20
	K		250	−1.58	0.30	12.5	45
Coastal	N	16.50	425	0.80	1.50	32.8	419
bermuda	P		43	−1.58	−0.65	27.3	28
	K		350	−1.58	0.30	19.5	63

$$N_u = \frac{A'}{1 + \exp(1.85 - 0.0100N)} \quad\quad [2.89]$$

$$N_c = N_{cm}\left[\frac{1 + \exp(1.20 - 0.0100N)}{1 + \exp(1.85 - 0.0100N)}\right] \quad\quad [2.90]$$

where $A = 10.60$ Mg ha^{-1}, $A' = 315$ kg ha^{-1}, and $N_{cm} = 29.7$ g kg^{-1} for common bermudagrass and $A = 16.50$ Mg ha^{-1}, $A' = 425$ kg ha^{-1}, and $N_{cm} = 25.8$ g kg^{-1} for coastal bermudagrass. The phase plot for N is shown in Fig. 2.17, where the curves are drawn from

Table 2.14 Estimated response of common and coastal bermudagrass to applied nutrients grown at Watkinsville, GA

Grass	N	P kg ha^{-1}	K	Y Mg ha^{-1}	N_u	P_u kg ha^{-1}	K_u	N_c	P_c g kg^{-1}	K_c
Common	0	0	0	2.25	40	5	27	17.8	2.2	12.0
bermuda	112	24	46	5.29	110	13	89	20.8	2.5	16.8
	224	48	92	8.09	201	22	165	24.8	2.7	20.4
	448	96	184	10.26	297	29	236	28.9	2.8	23.0
Coastal	0	0	0	3.52	53	7	38	15.1	2.0	10.8
bermuda	112	24	46	8.24	149	19	125	18.1	2.3	15.2
	224	48	92	12.60	272	31	231	21.6	2.5	18.3
	448	96	184	15.96	401	41	331	25.1	2.6	20.7

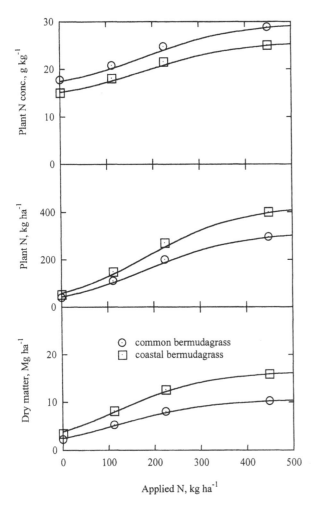

Figure 2.16 Dependence of yield and plant N uptake on applied N for coastal and common bermudagrass. Data from Adams et al. (1967). Curves drawn from Eqs. [2.88] through [2.90].

Common bermudagrass: $Y = \dfrac{22.4\,N_u}{350 + N_u}$ [2.91]

Coastal bermudagrass: $Y = \dfrac{34.9\,N_u}{475 + N_u}$ [2.92]

and the lines are drawn from

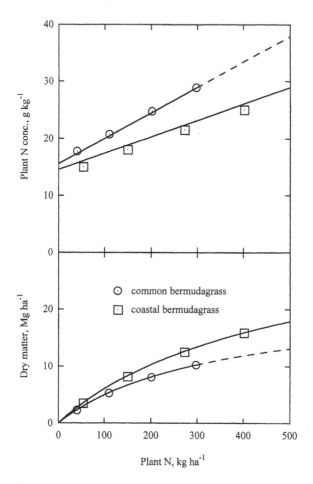

Figure 2.17   Dependence of dry matter on plant N uptake for coastal and common bermudagrass for experiment of Adams et al. (1967). Curves drawn from Eqs. [2.91] and [2.92]; lines drawn from Eqs. [2.93] and [2.94].

Common bermudagrass: $N_c = 15.6 + 0.0446 \, N_u$ [2.93]

Coastal bermudagrass: $N_c = 14.6 + 0.0287 \, N_u$ [2.94]

It is apparent that this approach works well for this system by simply changing values for b_n and b'_n. This is true because most of the dry matter response is due to applied N. It does not explain response of P_u and K_u.

2.4 WATER AVAILABILITY

Dependence of model parameters on soil moisture is further illustrated with data for perennial ryegrass (*L. perenne* L.) from a study in England by Morrison et al. (1980) and adapted from Whitehead (1995). Data are given in Table 2.15 for a dry site (Cambridge) and a wet site (Wales). Results are also shown in Fig. 2.18, where the curves are drawn from

$$Y = \frac{A}{1 + \exp(b - 0.0100N)} \qquad [2.95]$$

where $A = 6.25$ Mg ha^{-1} and $b = 1.20$ for the dry site, and $A = 15.1$ Mg ha^{-1} and $b = 2.32$ for the wet site. The corresponding semilog plot is shown in Fig. 2.19, where the lines are drawn from

$$\ln\left(\frac{A}{Y} - 1\right) = b - 0.0100N \qquad [2.96]$$

A common parameter c is indicated by the parallel lines. Clearly the wet site has greater production potential than the dry site at higher applied N levels, but the two are essentially the same with no applied N. These results are in agreement with findings of Overman et al. (1990b) for comparison of wet and dry years for production of coastal bermudagrass, for water availability and corn production (Reck and Overman, 1996), and for response of coastal bermudagrass and pensacola bahiagrass to rainfall (Overman and Evers, 1992).

Table 2.15 Perennial ryegrass response to applied nitrogen for dry and wet sites in England

Applied N kg ha^{-1}	Y, Mg ha^{-1}		$A/Y - 1$	
	Dry site	Wet site	Dry site	Wet site
0	1.4	1.3	3.46	10.6
150	3.4	4.8	0.838	2.15
300	5.6	9.7	0.116	·0.557
450	6.0	13.5	0.0417	0.117
600	6.2	14.7	0.00807	0.0272
A, Mg ha^{-1}			6.25	15.1
b			1.20	2.32
c, ha kg^{-1}			0.0100	0.0100
r			−0.9966	−0.99977

Data from Morrison et al. (1980) as adapted from Whitehead (1995).

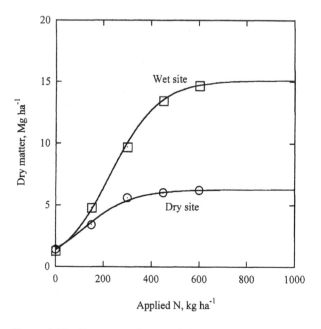

Figure 2.18 Response of perennial ryegrass to applied N for dry and wet sites in England. Curves are drawn from Eq. [2.95] with $A = 6.25$ Mg ha^{-1} and $b = 1.20$ for the dry site, and $A = 15.1$ Mg ha^{-1} and $b = 2.32$ for the wet site.

2.5 LEGUME/GRASS INTERACTION

Transfer of nitrogen from a legume (clover) to a grass (perennial ryegrass) is illustrated with other data from England, as given in Table 2.16 and shown in Fig. 2.20. The curve for grass sward only is drawn from

$$Y = \frac{11.1}{1 + \exp(1.18 - 0.0100N)} \qquad [2.97]$$

based on the semilog plot in Fig. 2.21, where the line is drawn from

$$\ln\left(\frac{11.1}{Y} - 1\right) = 1.18 - 0.0100N \qquad r = -0.9980 \qquad [2.98]$$

Since it appears that maximum yield for the grass-clover sward is the same as for the grass sward only, the curve for the grass-clover data is drawn from

$$Y = \frac{11.1}{1 + \exp(-0.15 - 0.0100N)} \qquad [2.99]$$

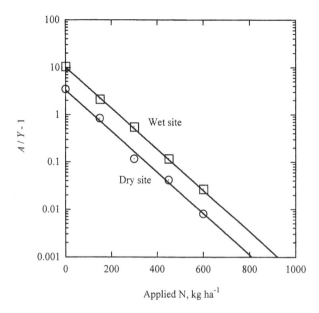

Figure 2.19 Semilog plot of dry matter response to applied N for dry and wet sites in England. Lines are drawn from Eq. [2.96] with $A = 6.25$ Mg ha^{-1} and $b = 1.20$ for the dry site, and $A = 15.1$ Mg ha^{-1} and $b = 2.32$ for the wet site.

Table 2.16 Perennial ryegrass response to applied nitrogen for grass and grass-clover swards in England

Applied N kg ha^{-1}	Y, Mg ha^{-1}		$A/Y - 1$
	Grass	Grass/Clover	Grass
0	2.7	7.1	3.11
100	5.0	8.2	1.22
200	7.6	9.2	0.460
300	9.4	10.1	0.181
400	10.6	10.9	0.0472
500	10.9	—	0.0183
600	11.0	—	0.00909
A, Mg ha^{-1}			11.1
b			1.18
c, ha kg^{-1}			0.0100
r			−0.9980

Data from Morrison et al. (unpublished) as adapted from Whitehead (1995).

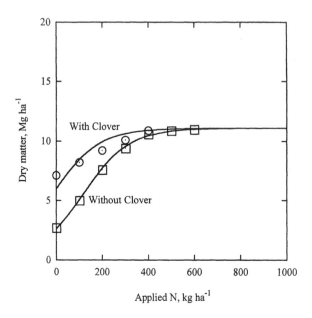

Figure 2.20 Response of perennial ryegrass to applied N for grass and grass-clover swards in England. Curves drawn from Eqs. [2.97] and [2.99].

to reflect the different intercept of $b = -0.15$. A common $c = 0.0100$ ha kg^{-1} is assumed. The line for the grass-clover in Fig. 2.21 is drawn from

$$\ln\left(\frac{11.1}{Y} - 1\right) = -0.15 - 0.0100N \qquad\qquad [2.100]$$

For $N = 0$, $Y = 6.0$ Mg ha^{-1} and the curve is estimated from Eq. [2.99]. The equivalent yield from Eq. [2.97] requires $N = 135$ kg ha^{-1}. We conclude that the clover supplies the equivalent of approximately 135 kg ha^{-1} of nitrogen. This is in the same range as 120 kg N ha^{-1} for bermudagrass and clover based on field studies from Jay, FL, Watkinsville, GA, and Eagle Lake, TX (Overman et al., 1992).

2.6 SUMMARY

We now return to question raised in Section 2.2 as to the ultimate limit on potential production. The quantity potential maximum dry matter yield Y_m was introduced in Eq. [2.16] and defined by Eq. [2.17]. It follows from Eq.

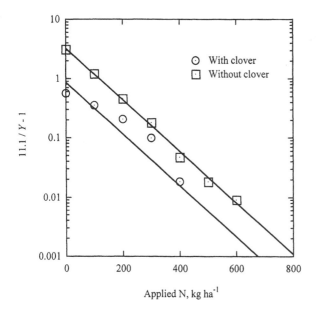

Figure 2.21 Semilog plot of dry matter response to applied N for grass and grass-clover swards in England. Lines drawn from Eqs. [2.98] and [2.100].

[2.17] that the ratio of actual maximum yield to potential maximum yield is given by

$$\frac{A}{Y_m} = 1 - \exp(-\Delta b) \qquad [2.101]$$

which indicates that this ratio is controlled by Δb. The significance of Δb in the physical system is defined by Eq. [2.7], which can be written as

$$\frac{N_{cl}}{N_{cm}} = \exp(-\Delta b) \qquad [2.102]$$

Since the lower and upper limits of plant nutrient concentration is characteristic of the plant species and cultivar, it follows that this ratio is controlled by the genetics of the plant. For a number of plant species the plant nutrient concentration ratio for nitrogen is approximately 1/2, from which we conclude that $A/Y_m \approx 1/2$.

It can be shown that the lower limit of plant nutrient concentration relates to the hyperbolic and logistic models by

$$\frac{K_n}{Y_m} = N_{cl} = N_{cm} \exp[-\Delta b] = \frac{A_n}{A} \exp[-(b_n - b)] \qquad [2.103]$$

The left side of Eq. [2.103] shows that N_{cl} relates to the two parameters (K_n and Y_m) of the hyperbolic equation (Eq. [2.16]), while the right side shows the relationship to the four parameters (A, A_n, b, and b_n) of the logistic equations (Eqs. [2.1] and [2.2]).

EXERCISES

2.1 Racine (1955) measured the response of pangolagrass (*Digitaria decumbens*) to applied N, P, and K on Leon fine sand (sandy, siliceous, thermic Aeric Haplaquods) at Gainesville, FL. Results for fixed P and K are given below.

    a. Plot dry matter Y, plant N uptake N_u, and plant N concentration N_c vs. applied N on linear graph paper.

    b. Calculate and draw Y vs. N on (a) from the logistic equation with $A = 16.00$ Mg ha^{-1}, $b = 1.50$, and $c = 0.0100$ ha kg^{-1}.

    c. Calculate and draw N_u vs. N on (a) from the logistic equation with $A_n = 410$ kg ha^{-1}, $b_n = 2.50$, and $c = 0.0100$ ha kg^{-1}.

    d. Calculate and draw N_c vs. N on (a) from the logistic equation with $N_{cm} = 26.5$ g kg^{-1}, $b = 1.50$, $b_n = 2.50$, and $c = 0.0100$ ha kg^{-1}.

    e. Plot Y and N_c vs. N_u on linear graph paper.

    f. Calculate the hyperbolic parameters Y_m and K_n.

    g. Calculate and plot Y and N_c vs. N_u on (e).

    h. Discuss the results. How well does the model describe the experimental results?

Dry matter and plant N response to applied N (1954, irrigated, applied $P = 25$ kg ha^{-1} and $K = 81$ kg ha^{-1}) for pangolagrass at Gainesville, FL

N kg ha^{-1}	Y Mg ha^{-1}	N_u kg ha^{-1}	N_c g kg^{-1}
0	2.16	22	10.2
21	2.83	31	11.0
43	3.30	37	11.3
86	5.96	79	13.2
173	9.83	145	14.8
259	12.52	225	18.0
432	15.05	327	21.7
604	15.80	393	24.9

Data adapted from Racine (1955).

2.2 Creel (1957) measured the response of coastal bermudagrass to applied N, P, and K on Leon fine sand (sandy, siliceous, thermic Aeric Haplaquods) at Gainesville, FL. Results averaged over the three levels of applied P and K are given below.

a. Plot dry matter Y, plant N uptake N_u, and plant N concentration N_c vs. applied N on linear graph paper.

b. Calculate and draw Y vs. N on (a) from the logistic equation with $A = 21.4$ Mg ha^{-1}, $b = 1.64$, and $c = 0.0047$ ha kg^{-1}.

c. Calculate and draw N_u vs. N on (a) from the logistic equation with $A_n = 587$ kg ha^{-1}, $b_n = 2.32$, and $c = 0.0047$ ha kg^{-1}.

d. Calculate and draw N_c vs. N on (a) from the logistic equation with $N_{cm} = 27.4$ g kg^{-1}, $b = 1.64$, $b_n = 2.32$, and $c = 0.0047$ ha kg^{-1}.

e. Plot Y and N_c vs. N_u on linear graph paper.

f. Calculate the hyperbolic parameters Y_m and K_n.

g. Calculate and plot Y and N_c vs. N_u on (e).

h. Discuss the results. How well does the model describe the experimental results?

Dry matter and plant N response to applied N (1955, averaged over P and K) for coastal bermudagrass at Gainesville, FL

N kg ha^{-1}	Y Mg ha^{-1}	N_u kg ha^{-1}	N_c g kg^{-1}
0	2.4	31	12.9
62	4.2	48	11.4
125	5.8	80	13.8
250	9.4	148	15.7
500	15.2	326	21.4
750	18.1	410	22.7
1250	20.4	558	27.4
1750	21.7	620	28.6

Data adapted from Creel (1957).

2.3 Reck (1992) analyzed response of corn to applied N following 3 years of fescue on a sandy loam soil at Watkinsville, GA (Carreker et al., 1977). The fescue sod was turned under after three years and corn grown over the next 4 years. Data are given in the table below.

a. Plot dry matter vs. applied N for each year on linear graph paper.

b. Calculate and plot Y vs. N on (a) for each year for the logistic equation using the parameters $A = 9.24$ Mg ha^{-1}; $c = 0.037$ ha kg^{-1}; and $b_1 = -1.66$, $b_2 = -1.03$, $b_3 = 0.02$, $b_4 = 0.48$ for years 1, 2, 3, and 4, respectively.

c. It is apparent from (a) that the residual nitrogen from the fescue sod diminishes with time. Estimate the effective N supplied from the fescue sod in the first year of corn.

d. Discuss the results. Which parameter in the model accounts for residual N? Does this seem reasonable to you?

Yield response to nitrogen for consecutive corn following three years of fescue at Watkinsville, GA (1967)

Applied N kg ha^{-1}	Dry matter Mg ha^{-1} Years after fescue			
	1	2	3	4
0	7.76	6.81	4.64	3.63
45	9.40	8.57	7.34	6.84
90	9.20	9.32	9.19	9.28
180	8.97	8.73	9.03	9.19

Data adapted from Carreker et al. (1977).

2.4 Rhoads and Olson (1995) measured response of the vegetable crops tomato and squash to organic fertilizer (mushroom compost) as given in Table 2.7.

a. Use linear graph paper to plot fresh wet of tomato Y_{tom}, vs. fresh weight of squash, Y_{sq}. Also plot the ratio Y_{sq}/Y_{tom} vs. Y_{sq} on the same graph.

b. Calculate the phase parameters

$$Y_m = \frac{A_{tom}}{1 - \exp(b_{tom} - b_{sq})} \qquad K' = \frac{A_{sq}}{\exp(b_{sq} - b_{tom})}$$

for the phase equation

$$Y_{tom} = \frac{Y_m Y_{sq}}{K' + Y_{sq}}$$

c. Plot the phase relationships Y_{tom} vs. Y_{sq} and Y_{sq}/Y_{tom} vs. Y_{sq} on (a).

d. Discuss the significance of these results.

2.5 Chowdhury and Rosario (1994) measured response of maize to applied N on a fine clayey (Ultic Haplustalf) soil at Los Banos, Phillipines.

a. Plot the data for total above ground plant on linear graph paper.
b. Draw model response on (a) from the equations

$$Y = \frac{8.00}{1 + \exp(0.30 - 0.0500N)}$$

$$N_u = \frac{70.0}{1 + \exp(0.75 - 0.0500N)}$$

$$N_c = 8.75 \left[\frac{1 + \exp(0.30 - 0.0500N)}{1 + \exp(0.75 - 0.0500N)} \right]$$

c. Plot Y and N_c vs. N_u on linear graph paper.
d. Calculate and plot the phase relations on (c).
e. Estimate the value of parameter A for the grain data for the equation

$$Y = \frac{A}{1 + \exp(0.30 - 0.0500N)}$$

f. Discuss the fit of the logistic model for total plant and grain.

Dependence of dry matter and plant N on applied N for maize at Los Banos, Phillippines

Component	N kg ha^{-1}	Y Mg ha^{-1}	N_u kg ha^{-1}	N_c g kg^{-1}
Total plant	0	3.43	24.8	7.23
	30	5.62	41.3	7.35
	60	7.66	63.8	8.33
	90	7.80	67.9	8.70
Grain	0	1.40		
	30	2.46		
	60	3.34		
	90	3.46		

Data from Chowdhury and Rosario (1994). Maize only.

2.6 Wilman and Mohamed (1980) measured dependence of dry matter and plant N of tall fescue on harvest interval and applied N for a rather stony soil with average depth of about 35 cm at Aberystwyth, Wales. Data are given below.
a. Plot dry matter Y, plant N uptake N_u, and plant N concentration N_c vs. applied N on linear graph paper.
b. Draw model response on (a) from the equations

$$Y = \frac{A}{1 + \exp(1.00 - 0.0075N)}$$

$$N_u = \frac{650}{1 + \exp(1.70 - 0.0075N)}$$

$$N_c = N_{cm} \left[\frac{1 + \exp(1.00 - 0.0075N)}{1 + \exp(1.70 - 0.0075N)} \right]$$

where $A(4) = 16.00$ and $A(8) = 20.00$ Mg ha^{-1} for harvest intervals of 4 and 8 wk.

c. Plot Y and N_c vs. N_u on linear graph paper.
d. Calculate the phase parameters Y_m and K_n and plot the appropriate equations on linear graph paper.
e. Discuss agreement between the data and model. Comment on the dependence of model parameter A on harvest interval.

Dry matter and plant N response to applied nitrogen and harvest interval for tall fescue at Aberystwyth, Wales

Harvest interval wk	Applied N kg ha^{-1}	Dry matter Mg ha^{-1}	Plant N uptake kg ha^{-1}	Plant N conc. g kg^{-1}
4	0	4.14	106	25.6
	131	8.29	230	27.7
	263	10.08	307	30.5
	394	13.91	489	35.2
	525	15.40	604	39.2
8	0	5.25	104	19.8
	131	12.16	258	21.2
	263	15.16	372	24.5
	394	17.37	515	29.6
	525	17.93	593	33.1

Data from Wilman and Mohamed (1980).

2.7 Parker et al. (1993) studied the response of tobacco to applied N on Ocilla sand (loamy, siliceous, thermic Aquic Arenic Paleudults) at Tifton, GA. Data are given in the table below.
a. Plot Y, N_u, and N_c vs. N on linear graph paper.
b. Draw the curves on (a) for the model

$$Y = \frac{3.40}{1 + \exp(-0.42 - 0.0200N)}$$

$$N_u = \frac{78.0}{1 + \exp(0.75 - 0.0200N)}$$

$$N_c = 23.0 \left[\frac{1 + \exp(-0.42 - 0.0200N)}{1 + \exp(0.75 - 0.0200N)} \right]$$

c. Plot Y and N_c vs. N_u on linear graph paper.
d. Draw the curve and the line on (c) for the model from

$$Y = \frac{4.93 N_u}{35.1 + N_u} \qquad N_c = 7.12 + 0.203 N_u$$

e. Discuss the analysis of variance given in the table for this system.
f. Use the standardized equation

$$A = Y[1 + \exp(-0.42 - 0.0200N)]$$

to estimate model parameter A for each year and each applied N for data in Table 3.
g. Use the standardized equation

$$A_n = N_u[1 + \exp(0.75 - 0.0200N)]$$

to estimate model parameter A_n for each year and each applied N for data in Table 4.
h. Use the data and equations to plot Y, N_u and N_c vs. N for each year on linear graph paper.
i. Discuss agreement between data and model for these data.

Table 1. Response of tobacco to applied N at Tifton, GA (1982–1984)

N kg ha^{-1}	Y Mg ha^{-1}	N_u kg ha^{-1}	N_c g kg^{-1}
0	2.01	24.7	12.3
28	2.54	36.0	14.2
56	2.87	45.8	16.0
84	3.01	54.4	18.1
140	3.22	69.7	21.6

Data adapted from Parker et al. (1993).

Table 2. Analysis of variance for model parameters for tobacco grown at Tifton, GA

Mode		Parameters estimated	df	Residual sum of squares	Mean sum of squares	F[1]
(1)	Common A, b, c	3	7	$1.27\ 10^{-1}$	$1.81\ 10^{-2}$	—
(2)	Individual A, b, c	6	4	$6.71\ 10^{-4}$	$1.68\ 10^{-4}$	—
(1)–(2)		—	3	$1.26\ 10^{-1}$	$4.20\ 10^{-2}$	250
(3)	Individual A, b	5	5	$1.97\ 10^{-3}$	$3.94\ 10^{-4}$	—
	Common c					
(3)–(2)		—	1	$1.30\ 10^{-3}$	$1.30\ 10^{-3}$	7.74

$F(3,4,95) = 6.59$; $F(1,4,95) = 7.71$.

Table 3. Estimates of model parameter A for tobacco at Tifton, GA (1982–1984)

N kg ha^{-1}	Y Mg ha^{-1}			A Mg ha^{-1}		
	1982	1983	1984	1982	1983	1984
0	2.64	1.90	1.48	4.37	3.15	2.45
28	3.10	2.52	2.01	4.26	3.47	2.76
56	3.42	2.81	2.38	4.15	3.41	2.89
84	3.58	2.91	2.55	4.02	3.27	2.86
140	3.84	2.97	2.84	3.99	3.09	2.95
avg	—	—	—	4.16	3.28	2.78
use	—	—	—	4.15	3.30	2.80

Yield data adapted from Parker et al. (1993).

Table 4. Estimates of model parameter A' for tobacco at Tifton, GA (1982–1984)

N kg ha^{-1}	N_u kg ha^{-1}			A_n kg ha^{-1}		
	1982	1983	1984	1982	1983	1984
0	34.3	22.2	17.6	106.9	69.2	54.9
28	44.0	33.3	30.8	97.2	73.6	68.0
56	56.8	40.5	40.2	96.0	68.5	68.0
84	68.0	49.8	45.4	94.8	69.4	63.3
140	86.8	56.7	65.6	98.0	64.0	74.0
avg	—	—	—	98.6	68.9	65.6
use	—	—	—	100.0	70.0	65.0

Plant N uptake data adapted from Parker et al. (1993).

2.8 Dasberg et al. (1983) reported response of "Shamouti" oranges (*Citrus sinemensis* L.) to applied N on Hamra sandy to sandy loam soil at Bet Dagan, Israel. Nutrients were applied through fertigation.
a. Plot fresh weight Y, fruit N removal N_u, and fruit N concentration N_c vs. applied N on linear graph paper.
b. Draw the curves on (a) for the model

$$Y = \frac{83}{1 + \exp(0.067 - 0.0115N)}$$

$$N_u = \frac{155}{1 + \exp(0.770 - 0.0115N)}$$

$$N_c = 1.87 \left[\frac{1 + \exp(0.067 - 0.0115N)}{1 + \exp(0.770 - 0.0115N)} \right]$$

c. Calculate the phase parameters Y_m and K_n.
d. Plot Y and N_c vs. N_u on linear graph paper.
e. Draw the curve and the line on (c) for the model from

$$Y = \frac{164 N_u}{152 + N_u} \qquad N_c = 0.927 + 0.00610 N_u$$

f. Calculate $N_{1/2}$ and $(N_{1/2})_n$.
g. Discuss the results.

Response of oranges to applied N at Bet Dagan,
Israel (1979–1981)

N kg ha^{-1}	Y Mg ha^{-1}	N_u kg ha^{-1}	N_c g kg^{-1}
100	61.7	91	1.47
170	71.7	120	1.67
310	80.4	146	1.82

Data adapted from Dasberg et al. (1983)

2.9 McConnell et al. (1993) reported response of irrigated cotton (*Gossypium hirsutum* L.) to applied nitrogen on Hebert silt loam (fine-silty, mixed, thermic Aeric Ochraqualf) at Rohwer, AR.
a. Plot the average of yield Y vs. applied nitrogen N on linear graph paper.
b. Plot the curve on (a) for the model

$$Y = \frac{A}{1 + \exp(b - cN)} = \frac{3.8}{1 + \exp(-1.20 - 0.0160N)}$$

c. Plot yield vs. applied N for each year on the same linear graph.
d. Plot the curves for the model

$$Y = \frac{A}{1 + \exp(-1.20 - 0.0160N)}$$

for each year with the parameters $A(1989) = 3.0$ Mg ha^{-1}, $A(1990)$ $= 3.9$ Mg ha^{-1}, $A(1991) = 4.4$ Mg ha^{-1}.

e. Discuss the results. Does the model describe the data adequately? Offer possible reasons for the variation of parameter A with year. Does the assumption of common b and c seem reasonable? What is the significance of the b parameter?

Response of seed cotton to applied nitrogen at Rohwer, AR

N kg ha^{-1}	Y Mg ha^{-1}			
	1989	1990	1991	Avg
0	1.86	3.05	3.05	2.69
56	2.60	3.52	3.60	3.24
112	3.02	3.79	4.35	3.75
168	2.79	3.75	4.35	3.66
224	3.02	3.77	4.48	3.76

Data from McConnell et al. (1993).

2.10 Bulman and Smith (1993) reported response of spring barley (*Hordeum vulgare* L.) to applied nitrogen on Bearbrook clay soil (fine, mixed, nonacid, frigid Humaquept) at Quebec, Canada.

a. Plot yield Y, plant N uptake N_u, and plant N concentration N_c vs. applied nitrogen N on linear graph paper for both grain and total components.

b. Plot the curves on (a) for the model

$$Y = \frac{A}{1 + \exp(-1.00 - 0.0210N)}$$

$$N_u = \frac{A_n}{1 + \exp(-0.35 - 0.0210N)}$$

$$N_c = N_{cm}\left[\frac{1 + \exp(-1.00 - 0.0210N)}{1 + \exp(-0.35 - 0.0210N)}\right]$$

where the parameters are $A(\text{total}) = 3.45$ g plant^{-1}, $A(\text{grain}) = 1.85$ g plant^{-1}, $A_n(\text{total}) = 55.0$ mg plant^{-1}, $A_n(\text{grain}) = 38.8$ mg plant^{-1}, N_{cm} (total) $= 15.9$ g kg^{-1}, and N_{cm} (grain) $= 21.0$ g kg^{-1}.

c. Show the phase plot (Y and N_c vs. N_u) on linear graph paper for both grain and total components.

d. Show that the phase equations for grain and total are given by

$$Y(\text{total}) = \frac{7.22N_u}{60.0 + N_u}$$

$$Y(\text{grain}) = \frac{3.87N_u}{42.4 + N_u}$$

$$N_c(\text{total}) = 8.3 + 0.138N_u$$

$$N_c(\text{grain}) = 11.0 + 0.258N_u$$

and draw these on (c).

e. Discuss the results.

Response of spring barley to applied nitrogen at Quebec, Canada

Plant component	N kg ha^{-1}	Y g plant^{-1}	N_u mg plant^{-1}	N_c g kg^{-1}
Total	0	2.58	31.1	12.1
	56	3.04	41.1	13.5
	112	3.37	49.5	14.7
	168	3.26	51.1	15.7
	224	3.41	58.1	17.0
Grain	0	1.37	24.4	17.8
	56	1.64	31.6	19.3
	112	1.79	35.8	20.0
	168	1.75	36.1	20.6
	224	1.78	38.6	21.7

Data adapted from Bulman and Smith (1993).

2.11 Reck (1992) analyzed response of corn to applied N following three years of Coastal bermudagrass on a sandy loam soil at Watkinsville, GA (Carreker et al., 1977). The bermudagrass sod was turned under after three years and corn grown over the next four years. Data are given in the table below.
   a. Plot dry matter vs. applied N for each year on linear graph paper.
   b. Calculate and plot Y vs. N on (a) for each year for the logistic equation using the parameters $A = 9.12$ Mg ha^{-1}; $c = 0.044$ ha kg^{-1}; and $b_1 = -2.69$, $b_2 = -0.50$, $b_3 = 0.08$, $b_4 = 0.15$ for years 1, 2, 3, and 4, respectively.
   c. It is apparent from (a) that the residual nitrogen from the bermudagrass sod diminishes with time. Estimate the effective N supplied from the bermudagrass sod in the first year of corn.
   d. Discuss the results. Which parameter in the model accounts for residual N? Does this seem reasonable to you?

Yield response to nitrogen for consecutive corn following 3 years of coastal bermudagrass at Watkinsville, GA (1967)

Applied N kg ha^{-1}	Dry matter Mg ha^{-1} Years after bermudagrass			
	1	2	3	4
0	8.54	5.68	4.42	4.27
45	8.56	8.67	7.56	7.72
90	9.19	9.16	9.21	9.54
180	9.11	8.94	9.06	8.86

Data from Carreker et al. (1977).

2.12 Reck (1992) analyzed response of corn cultivars to applied N from a study by Robertson et al. (1965) on Leon fine sand (sandy, siliceous, thermic Aeric Alaquod) at Quincy, FL.
   a. Plot dry matter yield Y vs. applied nitrogen N for the four cultivars on linear graph paper.
   b. Plot model curves on (a) from the equation

$$Y = \frac{A}{1 + \exp(0.76 - 0.011N)}$$

   where $A = 14.82$, 17.51, 18.77, 17.05 Mg ha^{-1} for Pioneer 309B, Dixie 18, Florida 200, and Coker 67 cultivars, respectively.

c. Discuss the results. Does the A parameter appear to account for variation among cultivars adequately?

Yield response of corn cultivars to applied N at Quincy, FL

Applied N kg ha^{-1}	Dry matter Mg ha^{-1}			
	Pioneer 309B	Dixie 18	Florida 200	Coker 67
0	4.8	5.8	5.6	5.3
168	11.2	12.9	14.8	13.6
336	13.7	16.8	17.2	16.0
672	15.3	17.7	19.2	16.9

Data from Robertson et al. (1965).

2.13 Reck (1992) analyzed response of continuous corn to applied N from a study by Carrekar et al. (1977) on soil of the Cecil series (fine, kaolinitic, thermic Typic Kanhapludults) at Watkinsville, GA
   a. Plot dry matter yield Y vs. applied nitrogen N for the 4 years on linear graph paper.
   b. Plot model curves on (a) from the equation

$$Y = \frac{A}{1 + \exp(1.15 - 0.033N)}$$

where $A(1965) = 5.03$, $A(1966) = 1.98$, $A(1967) = 8.60$, and $A(1968) = 4.68$ Mg ha^{-1}.
   c. Discuss the results. Does the A parameter appear to account for variation among years adequately?

Yield response of corn grain to applied N for continuous corn at Watkinsville, GA

Applied N kg ha^{-1}	Dry matter Mg ha^{-1}			
	1965	1966	1967	1968
0	1.46	0.85	1.76	0.83
45	3.77	1.83	4.60	3.07
90	4.04	1.29	7.39	3.84
180	4.71	1.85	8.89	4.70

Data from Carrekar et al. (1977).

2.14 Reck (1992) analyzed response of corn/rye rotation to applied N from a study by Carrekar et al. (1977) on soil of the Cecil series (fine, kaolinitic, thermic Typic Kanhapludults) at Watkinsville, GA.

   a. Plot dry matter yield Y vs. applied nitrogen N for the 4 years on linear graph paper.

   b. Plot model curves on (a) from the equation

$$Y = \frac{A}{1 + \exp(0.46 - 0.037N)}$$

   where $A(1965) = 5.49$, $A(1966) = 1.81$, $A(1967) = 7.81$, and $A(1968) = 4.30$ Mg ha^{-1}.

   c. Discuss the results. Does the A parameter appear to account for variation among years adequately?

Yield response of corn grain to applied N for corn/rye rotation at Watkinsville, GA

Applied N kg ha^{-1}	Dry matter Mg h^{-1}			
	1965	1966	1967	1968
0	2.79	1.26	2.49	1.35
45	4.81	1.40	5.62	3.65
90	4.77	1.52	7.37	3.86
180	5.15	1.75	8.29	4.34

Data from Carrekar et al. (1977).

2.15 We now wish to relate logistic model parameter A to seasonal rainfall for the corn study at Watkinsville, GA. Values are given in the table below.

   a. Plot model parameter A vs. rainfall W on linear graph paper.

   b. Perform linear regression on A vs. W to obtain the linear relationship

$$A = 0.257W - 0.855 \qquad r = 0.9934$$

   with the correlation coefficient $r = 0.9934$. Plot the regression line on (a).

   c. Discuss the results. Does the linear relationship appear to provide adequate dependence of parameter A on rainfall? What is the significance of the intercept value -0.855?

Dependence of logistic parameter A on rainfall (June and July) for corn grain yield at Watkinsville, GA

Year	Rainfall cm	Parameter A Mg ha^{-1} Continuous corn	corn/rye rotation
1965	23.9	5.03	5.49
1966	10.9	1.98	1.81
1967	35.3	8.60	7.81
1968	20.3	4.68	4.30

2.16 Morey et al. (1970) conducted an experiment on response of oats (*Avena sativa* L.) to applied N on Norfolk loamy sand (fine-loamy, kaolinitic, thermic Typic Kandiudults) at Tifton, GA. Data are given in the table below.

a. Plot dry matter Y, plant N uptake N_u, and plant N concentration N_c vs. applied N (N) on linear graph paper.

b. Draw the model curves on (a) from the equations

$$Y = \frac{1.90}{1 + \exp(0.30 - 0.0130N)}$$

$$N_u = \frac{75}{1 + \exp(1.30 - 0.0130N)}$$

$$N_c = 39.5 \left[\frac{1 + \exp(0.30 - 0.0130N)}{1 + \exp(1.30 - 0.0130N)} \right]$$

c. Plot the phase relations (Y and N_c vs. N_u) on linear graph paper.

d. Plot the model curves on (c) from the equations

$$Y = \frac{Y_m N_u}{K' + N_u} = \frac{3.00 N_u}{43.6 + N_u}$$

$$N_c = \frac{K_n}{Y_m} + \frac{1}{Y_m} N_u$$

e. Discuss the results. Does the model describe the data adequately?

Seasonal forage response to applied N for oats at Tifton, GA

Applied N kg ha^{-1}	Dry matter Mg ha^{-1}	Plant N removal kg ha^{-1}	Plant N concentration g kg^{-1}
0	0.81	14.8	18.3
45	1.15	27.7	24.1
90	1.23	32.3	26.3
135	1.62	58.3	36.0
180	1.49	51.6	34.6
224	1.78	63.2	35.5
336	2.00	76.4	38.2
448	1.74	68.7	39.5

Data adapted from Morey et al. (1970).

2.17 Blue and Graetz (1977) reported response of Pensacola bahiagrass to split applications of applied N on Myakka fine sand (sandy, siliceous, hyperthermic Aeric Alaquods) at Gainesville, FL. Treatments included 1, 2, 4, 8, and 16 applications during the growing season. Results are shown in the table below for dry matter yield Y, plant N uptake N_u, and plant N concentration N_c in response to applied nitrogen N for 1 and 16 applications.

a. Plot dry matter Y, plant N uptake N_u, and plant N concentration N_c vs. applied N (N) on linear graph paper for both 1 and 16 applications.

b. Draw the model curves on (a) from the equations

$$Y = \frac{A}{1 + \exp(1.40 - 0.0150N)}$$

$$N_u = \frac{A_n}{1 + \exp(1.90 - 0.0150N)}$$

$$N_c = N_{cm} \left[\frac{1 + \exp(1.40 - 0.0150N)}{1 + \exp(1.90 - 0.0150N)} \right]$$

where the parameters are given by: A (1 appl.) = 15.0 and A (16 appl.) = 17.0 Mg ha^{-1}; A_n (1 appl.) = 250 and A_n (16 appl.) = 280 kg ha^{-1}; N_{cm} (1 appl.) = 16.7 and N_{cm} (16 appl.) = 16.5 g kg^{-1}.

c. Plot the phase relations (Y and N_c vs. N_u) on linear graph paper.

d. Plot the model curves on (c) from the equations

$$Y = \frac{Y_m N_u}{K_n + N_u}$$

$$N_c = \frac{K_n}{Y_m} + \frac{1}{Y_m} N_u$$

where the parameters are given by: Y_m (1 appl.) = 38.1 and Y_m (16 appl.) = 43.2 Mg ha^{-1}; K_n (1 appl.) = 385 and K_n (16 appl.) = 432 kg ha^{-1}.

e. Discuss the results. Does the model describe the data adequately? Do you consider the difference between 1 and 16 applications very significant? Do you consider the range of applied N adequate to fully test this difference?

Seasonal dry matter yield, plant N removal, and plant N concentration for pensacola bahiagrass at Gainesville, FL

N kg ha^{-1}	Y Mg ha^{-1}	N_u kg ha^{-1}	N_c g kg^{-1}	Y Mg ha^{-1}	N_u kg ha^{-1}	N_c g kg^{-1}
		1 application			16 applications	
0	3.1	33.2	10.7	3.1	33.2	10.7
112	8.7	115	13.2	9.6	130	13.5
224	13.2	202	15.3	14.6	223	15.3

Data adapted from Blue and Graetz (1977).

2.18 Doss et al. (1960) studied response of three forage grasses to applied N and irrigation on Greenville fine sandy loam (fine, kaolinitic, thermic Rhodic Kandiudults) at Auburn, AL. The species of grasses was not specified. Results are given in the table below.
a. Plot dry matter yield Y vs. applied nitrogen N for both years and irrigation treatments on linear graph paper.
b. Draw the model curves on (a) from the logistic equation

$$Y = \frac{A}{1 + \exp(b - cN)} = \frac{A}{1 + \exp(1.35 - 0.0072N)}$$

where the A parameter has the values without irrigation of A(1956) = 6.0 and A(1957) = 21.0 Mg ha^{-1}; with irrigation of of A(1956) = 18.0 and A(1957) = 23.0 Mg ha^{-1}.
c. Discuss the results. Does it appear reasonable to use common values for parameters b and c between years and irrigation treatments? Comment on the significance of irrigation for the two years.

Yield response of forage grasses to applied N and irrigation at Auburn, AL

Irrigation	Applied N kg ha^{-1}	Dry matter yield Mg ha^{-1}	
		1956	1957
No	84	2.4	7.3
	168	4.2	11.8
	336	4.0	16.1
	672	4.0	20.2
Yes	84	5.6	7.7
	168	9.4	10.2
	336	12.8	16.8
	672	17.5	23.5

Data from Doss et al. (1960).

2.19 van Raij and van Diest (1979) reported response of six plant species grown in pots in a green house to different sources of phosphate. The grasses included wheat (*Triticum aestivum* cv. Kaspar), maize (*Zea mays* cv. Prior), soybean (*Glycine max* cv. Giesso), buckwheat (*Fagopyrum esculentum*), molasses grass (*Melinis munitiflora*), and paspalum grass (*Paspalum plicatulum*). Sources of phosphate were monocalcium phosphate, or triple superphosphate (S); calcined aluminum rock phosphate (CAP); and apatitic rock, or hyperphosphate (RP-H). Phosphorus was applied at 87 mg P kg^{-1} soil for all treatments. Data are listed in the table below. The high P concentrations in molasses and paspalum grasses for the RP-H source were unexplained.

a. Plot dry matter Y and plant P concentration P_c vs. plant P uptake P_u on linear graph paper for the six plant species and three sources of phosphate.

b. Perform linear regression of P_c vs. P_u to obtain the phase equations

$$P_c = \frac{K_p}{Y_m} + \frac{1}{Y_m}P_u = 0.743 + 0.0191P_u \qquad r = 0.9753$$

$$Y = \frac{Y_m P_u}{K_p + P_u} = \frac{52.3P_u}{38.8 + P_u}$$

Omit the two outliers $P_c = 2.12$ and 1.55 g kg^{-1}.

c. Draw the line and curve on the phase plot (a) from these equations.

d. Discuss the results. Does the hyperbolic model describe the correlation between dry matter and plant P uptake? Which source provides the greatest availability of P? Which grass responded best to applied P? Discuss the significance of the parameters Y_m and K_p.

Response of six plant species to three sources of phosphate in Wageningen, Netherlands

Species	P source	Dry matter g pot^{-1}	Plant P uptake mg pot^{-1}	Plant P concentration g kg^{-1}
Wheat	S	22.2	31.4	1.41
	CAP	14.8	14.7	0.99
	RP-H	4.3	4.17	0.97
Maize	S	19.5	17.7	0.91
	CAP	10.8	9.4	0.87
	RP-H	7.2	7.4	1.03
Soybean	S	12.1	11.7	0.97
	CAP	6.5	5.33	0.82
	RP-H	6.9	6.22	0.90
Buckwheat	S	33.7	64.8	1.92
	CAP	27.8	47.1	1.69
	RP-H	32.6	69.8	2.14
Molasses	S	19.9	25.9	1.30
grass	CAP	13.3	12.8	0.96
	RP-H	5.6	11.9	2.12
Paspalum	S	20.0	22.1	1.10
grass	CAP	17.7	16.2	0.92
	RP-H	6.2	9.6	1.55

Data adapted from van Raij and van Diest (1979, tables 2 and 3).

REFERENCES

Adams, W. E., M. Stelly, R. A. McCreery, H. D. Morris, and C. B. Elkins. 1966. Protein, P, and K composition of coastal bermudagrass and crimson clover. *J. Range Management* 19:301–305.

Adams, W. E., M. Stelly, H. D. Morris, and C. B. Elkins. 1967. A comparison of coastal and common bermudagrass [*Cynodon dactylon* (L.) Pers.] in the Piedmont region. II. Effect of fertilization and crimson clover (*Trifolium incarnatum*) on nitrogen, phosphorus, and potassium contents of the forage. *Agron. J.* 59:281–284.

Allen, Jr. L. H., E. C. Bisbal, W. J. Campbell, and K. J. Boote. 1990. Carbon dioxide effects on soybean developmental stages and expansive growth. *Soil and Crop Sci. Soc. Fla. Proc.* 49:124–131.

Asimov, I. 1966. *Understanding Physics: Motion, Sound, and Heat.* Barnes & Noble Books, New York.

Blue, W. G., and D. A. Graetz. 1977. The effect of split nitrogen applications on nitrogen uptake by Pensacola bahiagrass from an Aeric Haplaquod. *Soil Sci. Soc. Amer. J.* 41:927–930.

Bulman, P., and D. L. Smith. 1993. Accumulation and redistribution of dry matter and nitrogen by spring barley. *Agron. J.* 85:1114–1121.

Carreker, J. R., S. R. Wilkinson, A. P. Barnett, and J. E. Box. 1977. *Soil and water systems for sloping land.* ARS-S-160. U.S. Government Printing Office, Washington, DC.

Chowdhury, M. K., and E. L. Rosario. 1994. Comparison of nitrogen, phosphorus and potassium utilization efficiency in maize/mungbean intercropping. *J. Agric. Sci., Camb.* 122:193–199.

Creel, J. M., Jr. 1957. The effect of continuous high nitrogen fertilization on coastal bermudagrass. PhD dissertation. University of Florida, Gainesville, FL.

Dasberg, S., H. Bielorai, and J. Erner. 1983. Nitrogen fertigation of Shamouti oranges. *Plant and Soil* 75:41–49.

Doss, B. D., O. L. Bennett, D. A. Ashley, and L. E. Ensminger. 1960. Interrelation of nitrogen fertilization and irrigation of three forage grasses. *7th International Congress of Soil Science*, vol. IV. 63:496–502.

Hoffmann, B. 1972. *Einstein: Creator & Rebel.* New American Library, New York.

Johnson, F. H., H. Eyring, and B. J. Stover. 1974. *The Theory of Rate Processes in Biology and Medicine.* John Wiley & Sons, New York.

Kline, M. 1981. *Mathematics and the Physical World.* Dover Publications, New York.

Kramer, E. E. 1982. *The Nature and Growth of Modern Mathematics.* Princeton University Press, Princeton, NJ.

McConnell, J. S., W. H. Baker, D. M. Miller, B. S. Frizzell, and J. J. Varvil. 1993. Nitrogen fertilization of cotton cultivars of differing maturity. *Agron. J.* 85:1151–1156.

Morey, D. D., M. E. Walker, and W. H. Marchant. 1970. Influence of rates of nitrogen, phosphorus and potassium on forage and grain production of Suregrain oats in South Georgia. Georgia Agric. Exp. Stn. Research Report 64. University of Georgia, Athens, GA.

Morrison, J. H., M. V. Jackson, and P. E. Sparrow. 1980. The response of perennial ryegrass to fertilizer nitrogen in relation to climate and soil. Technical Report 27, Grassland Research Institute, Hurley, England.

Overman, A. R. 1995a. Rational basis for the logistic model for forage grasses. *J. Plant Nutr.* 18:995–1012.

Overman, A. R. 1995b. Coupling among applied, soil, root, and top components for forage crop production. *Commun. Soil Sci. Plant Anal.* 26:1179–1202.

Overman, A. R., and G. W. Evers. 1992. Estimation of yield and nitrogen removal by bermudagrass and bahiagrass. *Trans. Amer. Soc. Agric. Engr.* 35:207–210.

Overman, A. R., and S. R. Wilkinson. 1995. Extended logistic model of forage grass response to applied nitrogen, phosphorus, and potassium. *Trans. Amer. Soc. Agric. Engr.* 38:103–108.

Overman, A. R., and D. M. Wilson. 1999. Physiological control of forage grass yield and growth. In: *Crop Yield, Physiology and Processes*, pp. 443–473. D. L. Smith and C. Hamel (eds). Springer-Verlag, Berlin.

Overman, A. R., A. Dagan, F. G. Martin, and S. R. Wilkinson. 1991. A nitrogen-phosphorus-potassium model for forage yield of bermudagrass. *Agron. J.* 83:254–258.

Overman, A. R., G. W. Evers, and S. R. Wilkinson. 1995. Coupling of dry matter and nutrient accumulation in forage grass. *J. Plant Nutr.* 18:2629–2642.

Overman, A. R., F. G. Martin, and S. R. Wilkinson. 1990a. A logistic equation for yield response of forage grass to nitrogen. *Commun. Soil Sci. Plant Anal.* 21:595–609.

Overman, A. R., C. R. Neff, S. R. Wilkinson, and F. G. Martin. 1990b. Water, harvest interval, and applied nitrogen effects on forage yield of bermudagrass and bahiagrass. *Agron. J.* 82:1011–1016.

Overman, A. R., S. R. Wilkinson, and G. W. Evers. 1992. Yield response of bermudagrass and bahiagrass to applied nitrogen and overseeded clovers. *Agron. J.* 84:998–1001.

Overman, A. R., S. R. Wilkinson, and D. M. Wilson. 1994. An extended model of forage grass response to applied nitrogen. *Agron. J.* 86:617–620.

Parker, M. B., C. F. Douglas, S. H. Baker, M. G. Stephenson, T. P. Gaines, and J. D. Miles. 1993. Nitrogen, phosphorus, and potassium experiments on flue-cured tobacco. Georgia Agric. Exp. Sta. Res. Bull. 415. University of Georgia, Athens, GA.

Racine, R. G. 1955. The response of pangolagrass and coastal bermudagrass to high nitrogen fertilization. PhD dissertation. University of Florida, Gainesville, FL.

Ratkowsky, D. A. 1983. *Nonlinear Regression Modeling: A Unified Practical Approach*. Marcel Dekker, New York.

Reck, W. R. 1992. Logistic equation for use in estimating applied nitrogen effects on corn yields. M.S. Thesis. University of Florida, Gainesville, FL.

Reck, W. R. and A. R. Overman. 1996. Estimation of corn response to water and applied nitrogen. *J. Plant Nutr.* 19:201–214.

Rhoads, F. M. and S. M. Olson. 1995. Crop production with mushroom compost. *Soil and Crop Sci. Soc. Fla. Proc.* 54:53–57.

Robertson, W. K., L. C. Hammond, and L. G. Thompson Jr. 1965. Yield and nutrient removal by corn (*Zea mays* L.) for silage on two soil types as influenced by fertilizer, plant population and hybrids. *Soil Sci. Soc. Am. Proc.* 29:551–554.

Robinson, D. L., K. G. Wheat, N. L. Hubbert, M. S. Henderson, and H. J. Savoy, Jr. 1988. Dallisgrass yield, quality and nitrogen recovery responses to nitrogen and phosphorus fertilizers. *Commun. Soil Sci. Plant Anal.* 19:529–542.

Ruhla, C. 1993. *The Physics of Chance.* Oxford University Press, Oxford, England.

Silverman, M. P. 1998. *Waves and Grains: Reflections on Light and Learning.* Princeton University Press, Princeton NJ.

Stewart, I. 1989. *Does God Play Dice? The Mathematics of Chaos.* Basil Blackwell, Inc., Cambridge MA.

van Raij, B., and A. van Diest. 1979. Utilization of phosphate from different sources by six plant species. *Plant and Soil* 51:577–589.

Whitehead, D. C. 1995. *Grassland Nitrogen.* CAB International, Wallingford, UK.

Willcutts, J. F., A. R. Overman, G. J. Hochmuth, D. J. Cantliffe, and P. Soundy. 1998. A comparison of three mathematical models of response to applied nitrogen: A case study using lettuce. *HortScience* 33:833–836.

Williams, G. P. 1997. *Chaos Theory Tamed.* Joseph Henry Press. Washington, DC.

Wilman, D., and A. A. Mohamed. 1980. Response to nitrogen application and interval between harvests in five grasses-I. Dry-matter yield, nitrogen content and yield, numbers and weights of tillers, and proportion of crop fractions. *Fert. Res.* 1:245–263.

3

Growth Response Models

3.1 BACKGROUND

Dry matter and plant nutrient accumulation curves for forage grasses generally exhibit sigmoid shape with time. Growth rate is slow in the beginning as the plant develops, followed by a rapid phase, ending with decreasing rate as the plant matures. The middle phase often exhibits somewhat linear behavior. In this chapter we discuss several model attempts to capture the essential form of the growth response curves, culminating in the expanded growth model.

3.2 EMPIRICAL GROWTH MODEL

Overman (1984) used the probability function to describe accumulation of dry matter with time for coastal bermudagrass at Watkinsville, GA (Mays et al., 1980). The model takes the form

$$Y = \frac{A}{2}\left[1 + \mathrm{erf}\left(\frac{t-\mu}{\sqrt{2}\sigma}\right)\right]$$

[3.1]

where Y = accumulated dry matter, Mg ha^{-1}; t = calendar time since Jan. 1, wk; A = maximum accumulated dry matter, Mg ha^{-1}; μ = mean time of the dry matter distribution, wk; σ = time spread of the dry matter distribution, wk. The error function in Eq. [3.1] is defined by

$$\text{erf } x \doteq \frac{2}{\sqrt{\pi}} \int_0^x \exp(-u^2)\, du \tag{3.2}$$

which can be evaluated from mathematical tables (Abramowitz and Stegun, 1965). The model can also be normalized by defining

$$F = \frac{Y}{A} = \frac{1}{2}\left[1 + \text{erf}\left(\frac{t-\mu}{\sqrt{2}\sigma}\right)\right] \tag{3.3}$$

Equation [3.3] produces a straight line on probability paper. Estimates of μ and σ can be obtained by graphical means from data by noting that $\mu = t(F = 0.50)$ and $\sigma = [t(F = 0.84) - t(F = 0.16)]/2$. Another alternative is to linearize the model by rearranging Eq. [3.3] into the form

$$z = \text{erf}^{-1}(2F - 1) = -\frac{\mu}{\sqrt{2}\sigma} + \frac{t}{\sqrt{2}\sigma} \tag{3.4}$$

where erf^{-1} is the inverse of the error function. Linear regression is then performed on z vs. t to obtain $\mu/\sqrt{2}\sigma$ and $1/\sqrt{2}\sigma$ and therefore μ and σ. This latter procedure places heavy weight on very low and very high values of F.

The field experiment of Mays et al. (1980) consisted of a 2×2 factorial of two irrigation treatments (irrigated, nonirrigated) $\times$ two harvest intervals (4 wk, 6 wk). Yield data are listed in Tables 3.1 and 3.2. In this case ΔY = dry matter for a particular harvest, Mg ha^{-1}; Y = cumulative sum of dry matter, Mg ha^{-1}; A = total dry matter for the season, Mg ha^{-1}; $F = Y/A$. For example, for the nonirrigated treatment in Table 3.1 $A = 12.26$ Mg ha^{-1}. F is obtained by dividing each Y by this value. Since the F values for each time are very similar for irrigated and nonirrigated treatments, the two values are averaged. Values of z are calculated as follows. For $t = 28.7$ wk, for example, the average $F = 0.588$. Then $(2F - 1) = 0.176$. This means that $\text{erf}^{-1}(0.176) = z$. From math tables (Abramowitz and Stegun, 1965) we find that $z = 0.157$. In other words, $\text{erf}(0.157) = 0.176$. In a similar manner, for $t = 20.7$ wk, $F = 0.086$ and $(2F - 1) = -0.828$. Since $\text{erf}(-z) = -\text{erf}(+z)$, it follows from the tables that $z = -0.968$. Linear regression of z vs. t in Table 3.1 leads to

$$z = -3.47 + 0.124\, t \qquad r = 0.9952 \tag{3.5}$$

with a correlation coefficient of 0.9952. It follows from Eqs. [3.4] and [3.5] that $\sqrt{2}\sigma = 8.06$ wk and $\mu = 28.0$ wk. The curves in Fig. 3.1 are drawn from

Table 3.1 Dry matter accumulation with time for coastal bermudagrass with a harvest interval of 4 wk at Watkinsville, GA

t wk	ΔY Mg ha^{-1}	Y Mg ha^{-1}	F	ΔY Mg ha^{-1}	Y Mg ha^{-1}	F	F	z
	Nonirrigated			Irrigated			Avg	
16.7		0	0		0	0		
	1.01			1.40				
20.7		1.01	0.082		1.40	0.090	0.086	−0.968
	2.35			3.36				
24.7		3.36	0.274		4.76	0.307	0.290	−0.391
	3.86			4.34				
28.7		7.22	0.589		9.10	0.588	0.588	0.157
	3.08			3.30				
32.7		10.30	0.840		12.40	0.801	0.821	0.650
	0.95			1.79				
36.7		11.25	0.918		14.19	0.917	0.918	0.985
	1.01			1.29				
40.7		12.26	1		15.48	1		

Yield data adapted from Mays et al. (1980).

Table 3.2 Dry matter accumulation with time for coastal bermudagrass with a harvest interval of 6 wk grown at Watkinsville, GA

t wk	ΔY Mg ha^{-1}	Y Mg ha^{-1}	F	ΔY Mg ha^{-1}	Y Mg ha^{-1}	F	F	z
	Nonirrigated			Irrigated			Avg.	
16.7		0	0		0	0		
	2.48			4.62				
22.7		2.48	0.170		4.62	0.235	0.202	−0.590
	5.75			6.93				
28.7		8.23	0.564		11.55	0.587	0.576	−0.135
	4.96			5.67				
34.7		13.19	0.905		17.22	0.876	0.890	0.868
	1.39			2.44				
40.7		14.58	1		19.66	1		

Yield data adapted from Mays et al. (1980).

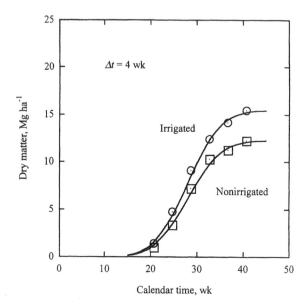

Figure 3.1 Dry matter accumulation with time by coastal bermudagrass for a harvest interval of 4 wks. Data from Mays et al. (1980). Curves drawn from Eq. [3.6] with $A = 15.48$ Mg ha^{-1} for irrigated and $A = 12.26$ Mg ha^{-1} for nonirrigated plots.

$$Y = \frac{A}{2}\left[1 + \mathrm{erf}\left(\frac{t - 28.0}{8.06}\right)\right]$$ [3.6]

where $A = 15.48$ Mg ha^{-1} for irrigated and $A = 12.26$ Mg ha^{-1} for non-irrigated treatments, respectively. The line in Fig. 3.2 is drawn from

$$F = \frac{1}{2}\left[1 + \mathrm{erf}\left(\frac{t - 28.0}{8.06}\right)\right]$$ [3.7]

A similar procedure for the harvest interval of 6 wk in Table 3.2 leads to

$$z = -3.35 + 0.122\,t \qquad r = 0.9999$$ [3.8]

with $\sqrt{2}\sigma = 8.20$ wk and $\mu = 27.6$ wk. The curves in Fig. 3.3 are drawn from

$$Y = \frac{A}{2}\left[1 + \mathrm{erf}\left(\frac{t - 27.6}{8.20}\right)\right]$$ [3.9]

where $A = 19.66$ Mg ha^{-1} for irrigated and $A = 14.58$ Mg ha^{-1} for non-irrigated treatments, respectively. The line in Fig. 3.4 is drawn from

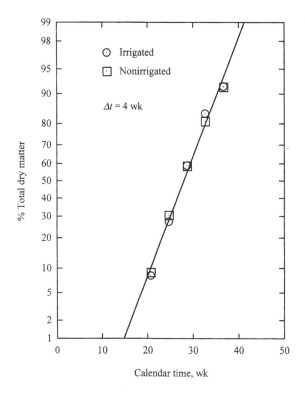

Figure 3.2 Probability plot of dry matter accumulation for coastal bermudagrass for harvest interval of 4 wk from study of Mays et al. (1980). Line drawn from Eq. [3.7].

$$F = \frac{1}{2}\left[1 + \operatorname{erf}\left(\frac{t - 27.6}{8.20}\right)\right] \qquad\qquad [3.10]$$

Since the parameters in Eqs. [3.7] and [3.10] are very similar for the two harvest intervals, linear regression on all of the z vs. t data gives

$$z = -3.43 + 0.123\,t \qquad r = 0.9959 \qquad\qquad [3.11]$$

with $\sqrt{2}\sigma = 8.13$ wk and $\mu = 27.8$ wk. These results suggest that the parameters μ and σ are independent of harvest interval and irrigation treatment, and that all of this dependence lies in the A parameter.

The meaning of the curves in Figs. 3.1 and 3.3 should be clarified. These do not provide information about growth of the grass between harvests, but instead relate accumulation for other harvest times that are on the same fixed harvest interval.

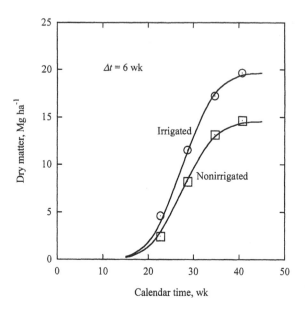

Figure 3.3 Dry matter accumulation with time by coastal bermudagrass for a harvest interval of 6 wk. Data from Mays et al. (1980). Curves drawn from Eq. [3.9] with $A = 19.66$ Mg ha^{-1} for irrigated and $A = 14.48$ Mg ha^{-1} for nonirrigated plots.

Mays et al. (1980) also reported on accumulation of plant P by coastal bermudagrass. Average harvest interval was 6.5 wk. Results are given in Table 3.3 and shown in Fig. 3.5. The linearized equation for plant P accumulation is given by

$$z = -3.50 + 0.136\,t \qquad r = 0.9984 \tag{3.12}$$

with $\sqrt{2}\sigma = 7.35$ wk and $\mu = 25.8$ wk. The curves in Fig. 3.5 are drawn from

$$P_u = \frac{A}{2}\left[1 + \mathrm{erf}\left(\frac{t - 25.8}{7.35}\right)\right] \tag{3.13}$$

with $A = 25.6$, 35.4, and 41.3 kg ha^{-1} for applied P $= 0$, 49, and 99 kg ha^{-1}, respectively. The line in the normalized graph (Fig. 3.6) is drawn from

$$F = \frac{1}{2}\left[1 + \mathrm{erf}\left(\frac{t - 25.8}{7.35}\right)\right] \tag{3.14}$$

It appears from this analysis that parameters μ and σ are independent of applied P and that all of this dependence is in linear parameter A.

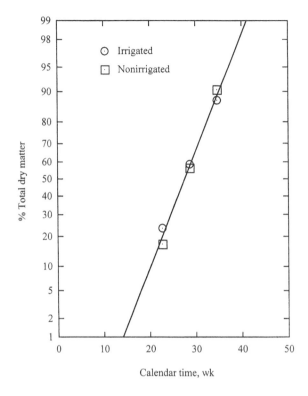

Figure 3.4 Probability plot of dry matter accumulation for coastal bermudagrass for harvest interval of 6 wk from study of Mays et al. (1980). Line drawn from Eq. [3.10].

3.3   EXTENDED EMPIRICAL GROWTH MODEL

In previous publications (Overman et al., 1988a; 1988b) the validity of the empirical model has been established. In this section we analyze a particular set of data in detail to link dry matter and plant N accumulation with time. The study was conducted by Day and Parker (1985) on coastal bermudagrass at Tifton, GA. Plant N uptake data were kindly provided to the author (J. L. Day, personal communication, 1987), and are listed in Table 3.4. Average harvest interval was 4.5 wk with applied N, P, K = 336, 36, 210 kg ha^{-1}. Normalized dry matter and plant N accumulation are shown in Fig. 3.7, where the line is drawn from

$$z = -3.32 + 0.133\,t \qquad r = 0.9965 \qquad\qquad [3.15]$$

Table 3.3 Phosphorus accumulation with time by coastal bermudagrass grown at Watkinsville, GA

t wk	ΔP_u kg ha^{-1}	P_u kg ha^{-1}	F	ΔP_u kg ha^{-1}	P_u kg ha^{-1}	F	ΔP_u kg ha^{-1}	P_u kg ha^{-1}	F	F
		$P = 0$ kg ha^{-1}			$P = 49$ kg ha^{-1}			$P = 99$ kg ha^{-1}		Avg.
15.4		0	0		0	0		0	0	
	5.0			7.5			10.1			
21.9		5.0	0.195		7.5	0.212		10.1	0.245	0.217
	9.8			13.3			14.5			
26.7		14.8	0.578		20.8	0.588		24.6	0.596	0.587
	7.7			10.2			11.0			
31.9		22.5	0.879		31.0	0.876		35.6	0.862	0.872
	3.1			4.4			5.7			
41.3		25.6	1		35.4	1		41.3	1	

Plant P data adapted from Mays et al. (1980).

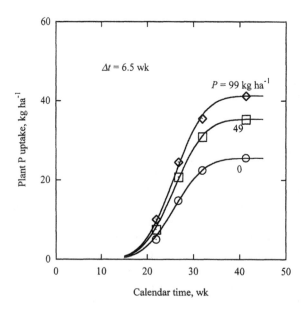

Figure 3.5 Plant P accumulation with time by coastal bermudagrass for a harvest interval of 6.5 wk. Data from Mays et al. (1980). Curves drawn from Eq. [3.13] with $A = 25.6$, 35.4, and 41.3 Mg ha^{-1} for applied $P = 0$, 49, and 99 kg ha^{-1}, respectively.

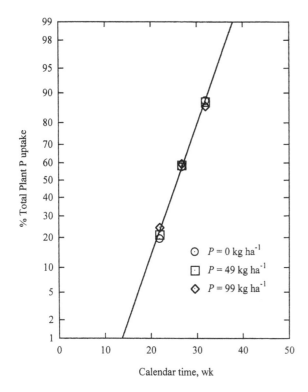

Figure 3.6 Probability plot of plant P accumulation for coastal bermudagrass for harvest interval of 6.5 wk from study of Mays et al. (1980). Line drawn from Eq. [3.14].

with $\sqrt{2}\sigma = 7.52$ wk and $\mu = 25.0$ wk. A plot of normalized plant N vs. normalized dry matter is shown in Fig. 3.8. It appears that the two are closely linked for a fixed harvest interval. For this case, average plant N concentration over the season is $N_c = N_u/Y = 270/16.00 = 16.9$ g kg^{-1}. It will be shown later that dry matter yield, plant N uptake, and plant N concentration are all functions of applied N.

3.4 PHENOMENOLOGICAL GROWTH MODEL

While the empirical growth model appears to fit field data rather well, it offers no insight into the physics of the system. The success of the empirical model suggests the presence of a Gaussian component in the system. Overman et al. (1989) wrote a growth rate equation (differential equation)

Table 3.4. Dry matter and plant N accumulation with time for coastal
bermudagrass grown at Tifton, GA

t wk	ΔY Mg ha^{-1}	Y Mg ha^{-1}	F	ΔN_u kg ha^{-1}	N_u kg ha^{-1}	F	F avg	z
15.4		0	0		0	0		
	2.60			47.6				
19.9		2.60	0.162		47.6	0.176	0.169	−0.677
	5.07			76.0				
24.3		7.67	0.479		123.6	0.458	0.468	−0.057
	3.99			70.6				
28.9		11.66	0.729		194.2	0.719	0.724	0.42
	3.57			61.7				
33.4		15.23	0.952		255.9	0.948	0.950	1.16
	—			—				
max		16.00	1		270.0	1		

Data from J. L. Day (personal communication, 1987).

as the product of a Gaussian environmental function and a linear intrinsic
growth function. Integration of the differential equation led to

$$\Delta Y_i = A \, \Delta Q_i \qquad [3.16]$$

where ΔY_i = incremental dry matter for growth interval i, Mg ha^{-1}; A =
yield factor, Mg ha^{-1}; and the growth quantifier for the ith growth interval
is defined by

$$\Delta Q_i = (1 - kx_i)[\mathrm{erf}(x_{i+1}) - \mathrm{erf}(x_i)] - \frac{k}{\sqrt{\pi}}[\exp(-x_{i+1}^2) - \exp(-x_i^2)]$$

$$[3.17]$$

where k = curvature parameter and the dimensionless time variable x_i is
defined by

$$x_i = \frac{t_i - \mu}{\sqrt{2}\sigma} \qquad [3.18]$$

with t_i = calendar time at the beginning of the ith growth interval, wk; μ =
mean time of the environmental function, wk; and σ = time spread of the
environmental function, wk. Now dry matter accumulation through the nth
growth interval can be calculated from the sum

$$Y_n = \sum_{i=1}^{n} \Delta Y_i = A \sum_{i=1}^{n} \Delta Q_i \qquad [3.19]$$

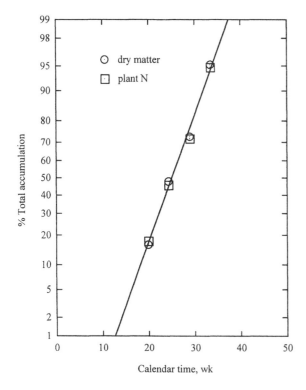

Figure 3.7   Probability plot of dry matter and plant N accumulation for coastal bermudagrass for harvest interval of 4.5 wk from study of Day and Parker (1985). Line drawn from Eq. [3.1] with $\mu = 25.0$ wk and $\sqrt{2}\sigma = 7.52$ wk.

If total dry matter yield for the season is defined as Y_t, then normalized yield for the nth growth interval is given as

$$F = \frac{Y_n}{Y_t} \qquad\qquad\qquad [3.20]$$

Overman et al. (1989) discussed the characteristics of this model and showed that F vs. t_i followed a straight line on probability paper. It was also shown that the model exhibited linear dependence of seasonal total yield Y_t on harvest interval for a fixed interval Δt. This was shown to be true for coastal bermudagrass for fixed harvest intervals through about 6 wk (Overman et al., 1989; 1990a). It was also shown that normalized yield data followed a straight line on probability paper, as predicted by the model.

We now examine a set of field data in considerable detail. Prine and Burton (1956) conducted a 2-year study at Tifton, GA with coastal bermu-

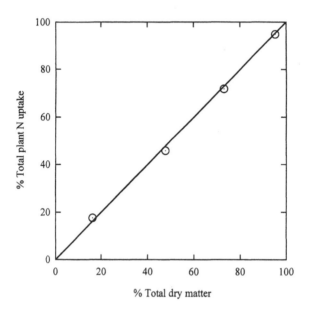

Figure 3.8 Normalized plant N vs. normalized dry matter for coastal bermudagrass from study of Day and Parker (1985). Line has 45 degree slope.

dagrass. Seasonal dry matter and plant N uptake were measured in response to harvest interval and applied N. Data are listed in Tables 3.5 through 3.7. While 1953 provided ideal moisture conditions, 1954 experienced a severe drought. For discussion purposes we will label these as 1953 (wet year) and 1954 (dry year). Model analysis of the dry matter data has been presented by Overman et al. (1990b), leading to the values for parameters A, b, and c listed in Table 3.8. Data for 1953 for harvest intervals of $\Delta t = 2$, 4, and 8 wk are shown in Fig. 3.9, where the lower curves for yields are drawn from

$$Y = \frac{A}{1 + \exp{(1.38 - 0.0077N)}} \qquad\qquad [3.21]$$

with appropriate A values taken from Table 3.8. Since plant N uptake appears essentially independent of harvest interval, this curve is drawn from

$$N_u = \frac{650}{1 + \exp{(2.10 - 0.0077N)}} \qquad\qquad [3.22]$$

Plant N concentration curves are drawn from

$$N_c = N_{cm}\left[\frac{1 + \exp(1.38 - 0.0077N)}{1 + \exp{(2.10 - 0.0077N)}}\right] \qquad\qquad [3.23]$$

Table 3.5 Seasonal dry matter yield dependence on harvest interval and applied nitrogen for coastal bermudagrass grown at Tifton, GA

| Year | Harvest interval wk | Dry matter, Mg ha^{-1} | | | | |
| | | Applied N, kg ha^{-1} | | | | |
		0	112	336	672	1008
1953	1	—	—	—	14.00	—
(wet year)	2	2.33	5.96	11.76	17.43	19.71
	3	3.33	8.91	13.64	19.24	20.47
	4	2.71	9.86	17.65	21.68	23.61
	6	4.35	12.77	21.79	28.11	30.11
	8	5.64	13.66	22.38	27.93	29.30
1954	1	—	—	—	5.33	—
(dry year)	2	0.76	2.69	6.83	7.84	8.62
	3	0.94	3.65	7.39	9.90	9.99
	4	1.08	4.55	9.45	11.13	11.49
	6	1.30	6.16	11.60	13.57	14.13
	8	1.93	6.45	12.23	15.86	16.22

Adapted from Prine and Burton (1956)

Table 3.6 Seasonal plant nitrogen uptake dependence on harvest interval and applied nitrogen for coastal bermudagrass grown at Tifton, GA

| Year | Harvest interval wk | Plant N uptake, kg ha^{-1} | | | | |
| | | Applied N, kg ha^{-1} | | | | |
		0	112	336	672	1008
1953	1	—	—	—	479	—
(wet year)	2	37	130	327	582	721
	3	51	184	363	579	682
	4	40	176	431	590	739
	6	53	158	392	621	738
	8	62	184	372	545	624
1954	1	—	—	—	217	—
(dry year)	2	17	70	224	299	320
	3	15	75	208	294	364
	4	19	80	233	323	344
	6	21	92	261	293	390
	8	26	100	223	325	417

Adapted from Prine and Burton (1956).

Table 3.7 Seasonal plant nitrogen concentration dependence on harvest interval and applied nitrogen for coastal bermudagrass grown at Tifton, GA

	Harvest interval wk	Plant N Concentration, g kg^{-1}				
		Applied N, kg ha^{-1}				
Year		0	112	336	672	1008
1953	1	—	—	—	34.2	—
(wet year)	2	16.0	21.8	27.8	33.4	36.6
	3	15.4	20.6	26.6	30.1	33.3
	4	14.8	17.9	24.4	27.2	31.3
	6	12.1	12.4	18.0	22.1	24.5
	8	11.0	13.5	16.6	19.5	21.3
1954	1	—	—	—	40.7	—
(dry year)	2	22.6	26.0	32.8	38.2	37.1
	3	16.0	20.5	28.1	29.7	36.4
	4	17.5	17.5	24.7	29.0	29.9
	6	16.4	14.9	22.5	21.6	27.6
	8	13.5	15.5	18.2	20.5	25.7

Adapted from Prine and Burton (1956).

Table 3.8 Model parameters for coastal bermudagrass grown at Tifton, GA

Year	Δt wk	A Mg ha^{-1}	A_n kg ha^{-1}	b	b_n	c ha kg^{-1}	Y_m Mg ha^{-1}	K_n kg ha^{-1}
1953	2	17.81	650	1.38	2.10	0.0077	34.69	616
	3	19.75					38.47	
	4	23.00					44.80	
	6	29.38					57.23	
	8	29.40					57.27	
	avg	23.87					46.50	
1954	2	8.34	340	1.38	2.10	0.0077	16.24	322
	3	9.88					19.25	
	4	11.59					22.58	
	6	14.28					27.82	
	8	16.14					31.44	
	avg	12.05					23.47	

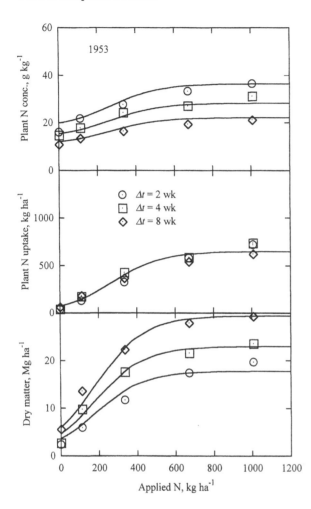

Figure 3.9 Dry matter and plant N response to applied N for coastal bermudagrass for 1953. Data from Prine and Burton (1956). Curves drawn from Eqs [3.21] through [3.23] with $A = 17.81$, 23.00, and 29.40 Mg ha^{-1} and $N_{cm} = 36.5$, 28.3, and 22.2 g kg^{-1} for harvest intervals of 2, 4, and 6 wk, respectively.

where $N_{cm} = 36.5$, 28.3, and 22.2 g kg^{-1} for harvest intervals of 2, 4, and 8 wk, respectively. Similar results are shown in Fig. 3.10 for 1954, where $A_n = 340$ kg ha^{-1} and $N_{cm} = 40.8$, 29.3, and 21.2 g kg^{-1} for harvest intervals of 2, 4, and 8 wk, respectively. The model describes the data rather well. Since it appears that the effects of water availability and harvest interval can be accounted for in the linear parameter A, it is statistically correct to average

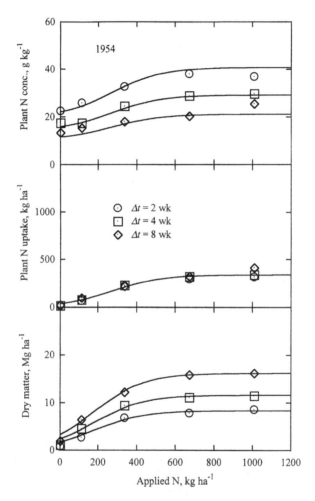

Figure 3.10 Dry matter and plant N response to applied N for coastal bermuda-grass for 1954. Data from Prine and Burton (1956). Curves drawn from Eqs. [3.23] through [3.25] with $A' = 340$ kg ha^{-1} and with $A = 8.34$, 11.59, and 16.14 intervals of 2, 4, and 6 wk, respectively.

dry matter over years and harvest intervals. These results are listed in Table 3.9 and shown in Fig. 3.11, where the yield curves are drawn from

$$Y = \frac{A}{1 + \exp(1.38 - 0.0077N)} \qquad [3.24]$$

with $A = 23.87$ Mg ha^{-1} for 1953 and $A = 12.05$ Mg ha^{-1} for 1954. Plant N uptake curves are drawn from

$$N_u = \frac{A_n}{1 + \exp(2.10 - 0.0077N)} \qquad [3.25]$$

with $A_n = 650$ kg ha^{-1} for 1953 and $A_n = 340$ kg ha^{-1} for 1954. Plant N concentration is then calculated from the ratio of Eqs. [3.25] to [3.24] with $N_{cm} = 27.2$ g kg^{-1} for 1953 and $N_{cm} = 28.2$ g kg^{-1} for 1954. Now Overman et al. (1994) have shown that seasonal accumulation of dry matter and plant N uptake can be related through the hyperbolic equation

$$Y = \frac{Y_m N_u}{K_n + N_u} \qquad [3.26]$$

which upon rearrangement leads to the linear relationship

$$N_c = \frac{N_u}{Y} = \frac{K_n}{Y_m} + \frac{N_u}{Y_m} \qquad [3.27]$$

where the new parameters Y_m and K_n are defined by

Table 3.9 Average response to applied nitrogen by coastal bermudagrass at Tifton, GA

Year	Applied N kg ha^{-1}	Dry matter Mg ha^{-1}	Plant N uptake kg ha^{-1}	Plant N conc. g kg^{-1}
1953	0	3.67	49	13.4
	112	10.23	166	16.2
	336	17.44	377	21.6
	672	22.88	583	25.5
	1008	24.64	700	28.4
1954	0	1.20	20	16.7
	112	4.70	83	17.7
	336	9.50	230	24.2
	672	11.66	307	26.3
	1008	12.09	367	30.4

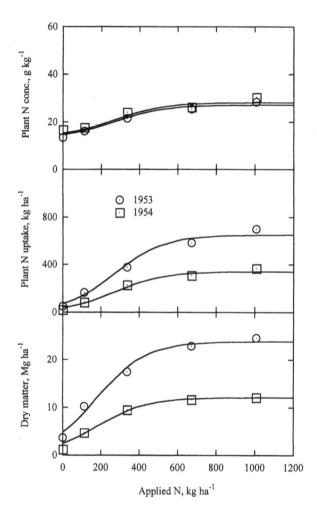

Figure 3.11 Dry matter and plant N response to applied N for coastal bermuda-grass for 1953 and 1954 averaged over harvest intervals. Data from Prine and Burton (1956). Curves drawn from Eqs. [3.23] through [3.25] with $A = 23.87$ and 12.05 Mg ha^{-1}, $A' = 650$ and 340 kg ha^{-1}, and $N_{cm} = 27.2$ and 28.2 g kg^{-1} for 1953 and 1954, respectively.

$$Y_m = \frac{A}{1 - \exp(\Delta b)} \qquad [3.28]$$

$$K_n = \frac{A_n}{\exp(\Delta b) - 1} \qquad [3.29]$$

with $\Delta b = b_n - b$. Since it appears that A_n is essentially independent of harvest interval for a given year, then it follows that K_n is also independent of harvest interval, and it becomes statistically acceptable to average over harvest intervals for each year and each applied N. Results from Table 3.9 are shown in Fig. 3.12, where the curves are drawn from

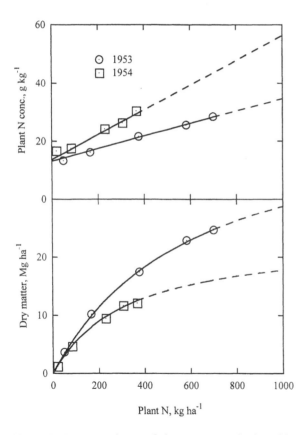

Figure 3.12 Dependence of dry matter and plant N concentration on plant N uptake for 1953 and 1954 from study of Prine and Burton (1956). Curves drawn from Eqs. [3.30] and [3.31] and lines from Eqs. [3.32] and [3.33] for 1953 and 1954, respectively.

$$1953: \quad Y = \frac{46.50\,N_u}{616 + N_u} \tag{3.30}$$

$$1954: \quad Y = \frac{23.47\,N_u}{322 + N_u} \tag{3.31}$$

and the lines from

$$1953: \quad N_c = 13.2 + 0.0215\,N_u \tag{3.32}$$

$$1954: \quad N_c = 13.7 + 0.0426\,N_u \tag{3.33}$$

It was pointed out above that one consequence of the phenomenological model is that seasonal total yield is a linear function of harvest interval. In turn, this should mean that maximum yield at high applied N should exhibit linear dependence on harvest interval. These results are shown in Fig. 3.13, where the lines are constructed from data for harvest intervals through 6 wk and are given by

$$1953: \quad A = 11.40 + 2.96\Delta t \qquad r = 0.9959 \tag{3.34}$$

$$1954: \quad A = 5.44 + 1.49\Delta t \qquad r = 0.9986 \tag{3.35}$$

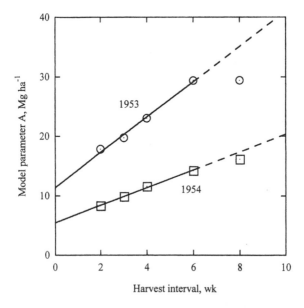

Figure 3.13 Dependence of model parameter A on harvest interval for coastal bermudagrass from study of Prine and Burton (1956). Lines drawn from Eqs. [3.34] and [3.35] for 1953 and 1954, respectively.

Apparently for harvest intervals beyond 6 wk there is an additional factor at work such that this model does not prove adequate. The model is modified to account for this effect in the next section. It should be noted that Eqs. [3.34] and [3.35] can be rewritten as

1953 : $A = 11.40(1 + 0.260\Delta t)$ [3.36]

1954 : $A = 5.44(1 + 0.274\Delta t)$ [3.37]

This suggests the more general relationship

$A = A_0(1 + 0.267\Delta t)$ [3.38]

where A_0 depends upon water availability. From this analysis, we conclude that for the coastal bermudagrass at Tifton, GA yields for 1954 were approximately half those for 1953 for all harvest intervals and at all applied N.

3.5 EXPANDED GROWTH MODEL

The phenomenological model discussed above proved to have a limited range of applicability for perennial grasses. Furthermore, it does not apply for annuals such as corn since the intrinsic growth function allows unlimited growth. To correct this deficiency, Overman (1998) expanded the intrinsic growth function to restrict growth. This model is described by

$$\Delta Y_i = A \, \Delta Q_i$$ [3.39]

where ΔY_i = accumulated dry matter for ith growth interval, Mg ha^{-1}; A = yield factor, Mg ha^{-1}; and ΔQ_i = growth quantifier now defined by

$$\Delta Q_i = \exp(\sqrt{2}\sigma c x_i)\left\{(1 - kx_i)(\text{erf}\,x - \text{erf}\,x_i) - \frac{k}{\sqrt{\pi}}[\exp(-x^2) - \exp(-x_i^2)]\right\}$$ [3.40]

where c = exponential decay coefficient, k = curvature factor, and x = dimensionless time defined by

$$x = \frac{t - \mu}{\sqrt{2}\sigma} + \frac{\sqrt{2}\sigma c}{2}$$ [3.41]

where μ = calendar time to the mean of the environmental distribution, wk; σ = time spread of the environmental distribution, wk. For a perennial grass cumulative yield through the nth harvest is given by

$$Y_n = \sum_{i=1}^{n} \Delta Y_i = A \sum_{i=1}^{n} \Delta Q_i \qquad\qquad [3.42]$$

If total dry matter yield for the season is defined as Y_t, then normalized yield for the nth growth interval is given as

$$F = \frac{Y_n}{Y_t} \qquad\qquad [3.43]$$

Overman (1998) showed that for a fixed harvest interval, Eq. [3.43] for F vs. t produces a straight line on probability paper, which can be described by

$$F = \frac{1}{2}\left[1 + \text{erf}\left(\frac{t - \mu}{\sqrt{2}\sigma}\right)\right] \qquad\qquad [3.44]$$

with the same mean and standard deviation as the environmental distribution. A further consequence of this model is that, for a fixed harvest interval Δt, seasonal dry matter yield is given by the linear-exponential equation

$$Y_t = (\alpha + \beta\,\Delta t)\,\exp\,(-\gamma\,\Delta t) \qquad\qquad [3.45]$$

where α, β, γ are regression coefficients. Note that for $\gamma = 0$ this model reduces to the phenomenological model. It will be further assumed that plant N uptake follows similar dependence on harvest interval and is given by

$$N_u = (\alpha_n + \beta_n\,\Delta t)\exp\,(-\gamma\,\Delta t) \qquad\qquad [3.46]$$

where α_n and β_n are regression coefficients and with same exponential coefficient as for dry matter. Plant N concentration is then defined as the ratio of Eqs. [3.46] to [345]:

$$N_c = \frac{\alpha_n + \beta_n\,\Delta t}{\alpha + \beta\,\Delta t} \qquad\qquad [3.47]$$

Equations [3.45] and [3.46] can be standardized by multiplying through by the exponential term to give

$$Y_t^* = Y_t\exp\,(\gamma\Delta t) = \alpha + \beta\Delta t \qquad\qquad [3.48]$$

$$N_u^* = N_u\exp\,(\gamma\,\Delta t) = \alpha_n + \beta_n\,\Delta t \qquad\qquad [3.49]$$

Parameter γ is chosen to give the best linear relationships for Y_t^* vs. Δt and N_u^* vs. Δt.

We now apply the expanded growth model to data for a perennial grass. Burton et al. (1963) reported data for coastal bermudagrass grown at Tifton, GA. Results are listed in Table 3.10 and shown in Fig. 3.15, where the lines are given by

Table 3.10 Dry matter and plant N dependence on harvest interval for coastal bermudagrass grown at Tifton, GA

Δt wk	Y_t Mg ha^{-1}	N_u kg ha^{-1}	N_c g kg^{-1}	$Y_t \exp(0.077\,\Delta t)$ Mg ha^{-1}	$N_u \exp(0.077\,\Delta t)$ kg ha^{-1}
3	15.2	438	28.8	19.1	552
4	16.2	415	25.6	22.0	565
5	17.8	417	23.4	26.2	613
6	19.9	411	20.6	31.6	652
8	19.9	340	17.1	36.8	630
12	20.1	289	14.4	50.6	728
24	14.6	198	13.6	92.7	1257

Yield and plant N data adapted from Burton et al. (1963).

$$Y_t^* = Y_t \exp(0.077\Delta t) = 8.91 + 3.49\Delta t \qquad [3.50]$$

$$N_u^* = N_u \exp(0.077\Delta t) = 422 + 33.0\Delta t \qquad [3.51]$$

Curves in Fig. 3.14 are drawn from

$$Y_t = (8.91 + 3.49\Delta t)\exp(-0.077\Delta t) \qquad [3.52]$$

$$N_u = (422 + 33.0\Delta t)\exp(-0.077\Delta t) \qquad [3.53]$$

$$N_c = \frac{422 + 33.0\Delta t}{8.91 + 3.49\Delta t} \qquad [3.54]$$

The expanded growth model appears to describe field data rather well.

Our analysis above has focused on clipping data and how it relates to applied N, harvest interval, and available water. The question still remains on whether the model also simulates growth between harvests. Fortunately, data of this type have been reported. Villanueva (1974) measured growth of coastal bermudagrass at College Station, TX. Plant samples were collected weekly during the second 8-wk growth interval at applied N levels of $N = 120$, 300, and 600 kg ha^{-1}. Results are listed in Table 3.11 and plotted in Fig. 3.16. The challenge now is to perform the simulation calculations with the model. Based on experience, parameter values are chosen as follows: $t_i = 23.0$ wk, $\mu = 26.0$ wk, $\sqrt{2}\sigma = 8.0$ wk, $c = 0.2$ wk^{-1}, $k = 5$. Time of initiation of significant growth t_i is chosen to match the growth curve. The dimensionless time variable, x, is now given by

$$x = \frac{t - \mu}{\sqrt{2}\sigma} + \frac{\sqrt{2}\sigma c}{2} = \frac{t - 26.0}{8.0} + 0.8 = \frac{t - 19.6}{8.0} \qquad [3.55]$$

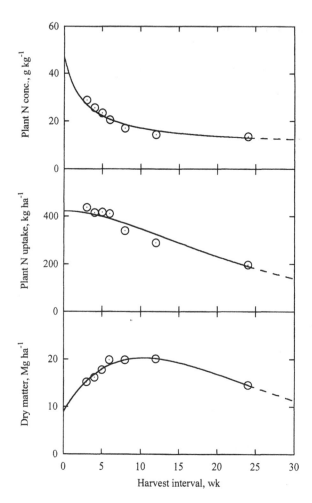

Figure 3.14 Dependence of dry matter and plant N uptake on harvest interval for coastal bermudagrass. Data from Burton et al. (1963). Curves drawn from Eqs. [3.52] through [3.54].

while the growth quantifier becomes

$$\Delta Q_i = \exp\left(\sqrt{2}\sigma c x_i\right)\left\{(1 - kx_i)(\text{erf } x - \text{erf } x_i) - \frac{k}{\sqrt{\pi}}\left[\exp(-x^2) - \exp(-x_i^2)\right]\right\}$$

$$= 1.974\left\{-1.125(\text{erf } x - 0.452) - 2.821\left[\exp(-x^2) - 0.835\right]\right\} \qquad [3.56]$$

Values of the error function are obtained from Abramowitz and Stegun (1965). The second match point is chosen at $t = 29$ wk with dry matter

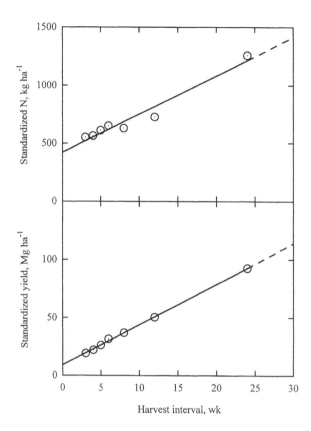

Figure 3.15 Dependence of standardized dry matter and standardized plant N on harvest interval for study of Burton et al. (1963). Lines drawn from Eqs. [3.50] and [3.51].

accumulation estimated to be 5.85 Mg ha^{-1} for $N = 600$ kg ha^{-1} and the growth quantifier at 2.25, so that

$$N = 600 \text{ kg ha}^{-1}: \quad \Delta Y = (5.85/2.25)\Delta Q = 2.60\Delta Q \qquad [3.57]$$

Estimates of dry matter accumulation in Table 3.12 and Fig. 3.16 are calculated from Eq. [3.57]. Simulations at the other two applied N levels are calculated from

$$N = 300 \text{ kg ha}^{-1}: \quad \Delta Y = (4.70/2.25)\Delta Q = 2.09\Delta Q \qquad [3.58]$$

$$N = 120 \text{ kg ha}^{-1}: \quad \Delta Y = (3.50/2.25)\Delta Q = 1.55\Delta Q \qquad [3.59]$$

Table 3.11 Dry matter accumulation during 2nd 8-wk interval by coastal bermudagrass at College Station, TX

| t | ΔY, Mg ha^{-1} | | |
| | N, kg ha^{-1} | | |
wk	120	300	600
21.5	0	0	0
22.5	0.10	0.07	0.04
23.5	0.63	0.48	0.46
24.5	1.27	1.60	1.60
25.5	1.48	2.00	2.44
26.5	1.59	2.41	2.94
27.5	2.06	2.77	3.83
28.5	3.26	4.21	5.27
29.5	3.76	5.10	6.33

Data from Villanueva (1974).

The shapes of the simulation curves appear to be in reasonable agreement with the data. It is now possible to simulate the growth curves for all of the 8-wk intervals over the year for applied N of 600 kg ha^{-1}. Calculations are listed in Tables 3.12 through 3.16 with the equations listed below each table. Results are summarized in Fig. 3.17. The seasonal effect is apparent, with slow growth early in the year followed by rapid growth in midsummer and followed by slow growth again late in the year. The Gaussian envelope of the terminus points of the curves is also apparent, which is further confirmed in Table 3.17 and Fig. 3.18 for cumulative dry matter over the season.

The expanded growth model can also be applied to an annual crop. Karlen et al. (1987) reported growth response of corn at Florence, SC. Data are given in Table 3.18. Response curves for dry matter and plant N are shown in Fig. 3.19. Dry matter accumulation is estimated from the model with the following parameters: $t_i = 19.0$ wk, $\mu = 26.0$ wk, $\sqrt{2}\sigma = 8.0$ wk, $c = 0.2$ wk^{-1}, $k = 5$. Dimensionless time is defined by Eq. [3.41] and becomes

$$x = \frac{t - 19.6}{8.0} \qquad [3.60]$$

and the growth quantifier, defined by Eq. [3.40], becomes

$$Q = 0.887 \left\{ 1.375 \left(\text{erf } x + 0.085 \right) - 2.821 \left[\exp \left(-x^2 \right) - 0.9944 \right] \right\} \qquad [3.61]$$

In this case the Δ is dropped for convenience. Simulation calculations are given in Table 3.19. Dry matter accumulation is estimated from

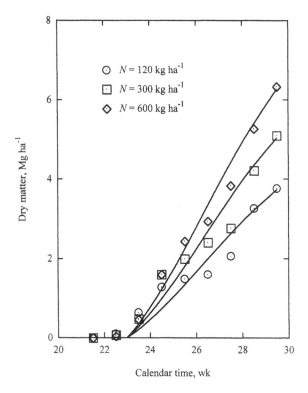

Figure 3.16 Dry matter accumulation with time by coastal bermudagrass for applied N = 120, 300, and 600 kg ha^{-1}. Data from Villanueva (1974). Curves drawn from Eqs. [3.55] through [3.59].

$$Y = (24.0/3.27)Q = 7.34Q \qquad [3.62]$$

where the match point at 30 wk is used for calibration. A phase plot of plant N accumulation vs. dry matter accumulation is shown in Fig. 3.20. The data appear to follow the hyperbolic relationship

$$N_u = \frac{N_{um} Y}{K_y + Y} \qquad [3.63]$$

where N_{um} = maximum plant N accumulation, kg ha^{-1}; K_y = yield response coefficient, Mg ha^{-1}. Equation [3.63] can be rearranged to the form

$$\frac{Y}{N_u} = \frac{K_y}{N_{um}} + \frac{Y}{N_{um}} \qquad [3.64]$$

The plot of Y/N_u vs. Y in Fig. 3.20 follows the straight line

Table 3.12 Model estimates of growth of coastal bermudagrass at College Station, TX for beginning time of 21.5 wk and applied N of 600 kg ha^{-1}

t wk	x	erf x	$\exp(x^2)$	ΔQ	ΔY Mg ha^{-1}
21.5	—	—	—	—	—
23.0	0.425	0.452	0.835	0	0
24.0	0.550	0.563	0.739	0.288	0.75
25.0	0.675	0.660	0.634	0.657	1.71
26.0	0.800	0.742	0.527	1.07	2.78
27.0	0.925	0.810	0.425	1.49	3.87
28.0	1.050	0.862	0.332	1.89	4.91
29.0	1.175	0.903	0.251	2.25	5.85
29.5	1.2375	0.919	0.216	2.41	6.27
30.0	1.300	0.934	0.185	2.55	6.63

Assume: $\quad t_i = 23.0\,\text{wk},\ \mu = 26.0\,\text{wk},\ \sqrt{2}\sigma = 8.0\,\text{wk},\ c = 0.2\,\text{wk}^{-1},\ k = 5$

Then: $\quad x = \dfrac{t - \mu}{\sqrt{2}\sigma} + \dfrac{\sqrt{2}\sigma c}{2} = \dfrac{t - 26.0}{8.0} + 0.8 = \dfrac{t - 19.6}{8.0}$

$$\Delta Q = \exp\left(\sqrt{2}\sigma c x_i\right)\left\{(1 - kx_i)[\text{erf } x - \text{erf } x_i] - \frac{k}{\sqrt{\pi}}[\exp(-x^2) - \exp(-x_i^2)]\right\}$$

$$= 1.974\left\{-1.125\,[\text{erf } x - 0.452] - 2.821[\exp(-x^2) - 0.835]\right\}$$

$$\Delta Y = \left(\frac{5.85}{2.25}\right)\Delta Q = 2.60\ \Delta Q$$

$$\frac{Y}{N_u} = 0.0250 + 0.00368\ Y \qquad r = 0.9975 \qquad\qquad [3.65]$$

with a correlation coefficient of 0.9975. It follows that the curve for plant N accumulation vs. dry matter accumulation in Fig. 3.20 is described by

$$N_u = \frac{270\ Y}{6.8 + Y} \qquad\qquad [3.66]$$

Estimates of plant N accumulation in Table 3.19 are made by substituting estimates of dry matter accumulation into Eq. [3.66]. Plant N concentration, $N_c = N_u/Y$, is related to dry matter by

$$N_c = \frac{270}{6.8 + Y} \qquad\qquad [3.67]$$

Table 3.13 Model estimates of growth of coastal bermudagrass at College Station, TX for beginning time of 5.5 wk and applied N of 600 kg ha^{-1}

t wk	x	erf x	$\exp(-x^2)$	ΔQ	ΔY Mg ha^{-1}
5.5	—	—	—	—	—
7.0	−1.575	−0.974	0.0837	0	0
8.0	−1.450	−0.960	0.122	0.001	0.00
9.0	−1.325	−0.939	0.173	0.005	0.01
10.0	−1.200	−0.910	0.237	0.011	0.03
11.0	−1.075	−0.872	0.315	0.020	0.05
12.0	−0.950	−0.821	0.406	0.036	0.09
13.0	−0.825	−0.757	0.506	0.059	0.15
13.5	−0.7625	−0.720	0.559	0.074	0.19
14.0	−0.700	−0.678	0.613	0.091	0.24

Assume: $t_i = 7.0\,\text{wk}$, $\mu = 26.0\,\text{wk}$, $\sqrt{2}\sigma = 8.0\,\text{wk}$, $c = 0.2\,\text{wk}^{-1}$, $k = 5$

Then: $$x = \frac{t - \mu}{\sqrt{2}\sigma} + \frac{\sqrt{2}\sigma c}{2} = \frac{t - 26.0}{8.0} + 0.8 = \frac{t - 19.6}{8.0}$$

$$\Delta Q = \exp\left(\sqrt{2}\sigma c x_i\right)\left\{(1 - kx_i)[\text{erf } x - \text{erf } x_i] - \frac{k}{\sqrt{\pi}}[\exp(-x^2) - \exp(-x_i^2)]\right\}$$

$$= 0.0805\left\{8.875\,[\text{erf } x + 0.974] - 2.821[\exp(-x^2) - 0.0837]\right\}$$

$$\Delta Y = 2.60\,\Delta Q$$

as shown in Table 3.19 and Fig. 3.20. The simulation model matches the form of the data rather well. A similar procedure for plant P accumulation P_u and plant P concentration P_c leads to the equations

$$\frac{Y}{P_u} = 0.269 + 0.0172\,Y \qquad r = 0.9965 \tag{3.68}$$

$$P_u = \frac{58.1\,Y}{15.6 + Y} \tag{3.69}$$

with the results shown in Fig. 3.21. Finally, this procedure leads to

$$\frac{Y}{K_u} = 0.00635 + 0.00316\,Y \qquad r = 0.9894 \tag{3.70}$$

$$K_u = \frac{317\,Y}{2.01 + Y} \tag{3.71}$$

Table 3.14 Model estimates of growth of coastal bermudagrass at College
Station, TX for beginning time of 13.5 wk and applied N of 600 kg ha^{-1}

t wk	x	erf x	$\exp(-x^2)$	ΔQ	ΔY Mg ha^{-1}
13.5	—	—	—	—	—
15.0	0.575	−0.586	0.718	0	0
16.0	0.450	−0.475	0.817	0.060	0.16
17.0	0.325	−0.354	0.900	0.153	0.40
18.0	0.200	−0.223	0.961	0.287	0.75
19.0	0.075	−0.085	0.9944	0.462	1.20
20.0	+0.500	+0.056	0.9975	0.676	1.76
21.0	+0.175	+0.195	0.970	0.922	2.40
21.5	+0.2375	+0.263	0.945	1.055	2.74
22.0	+0.300	+0.329	0.914	1.191	3.10

Assume: $t_i = 15.0$ wk, $\mu = 26.0$ wk, $\sqrt{2}\sigma = 8.0$ wk, $c = 0.2$ wk^{-1}, $k = 5$

Then: $x = \dfrac{t - \mu}{\sqrt{2}\sigma} + \dfrac{\sqrt{2}\sigma c}{2} = \dfrac{t - 26.0}{8.0} + 0.8 = \dfrac{t - 19.6}{8.0}$

$$\Delta Q = \exp\left(\sqrt{2}\sigma c x_i\right)\left\{(1 - kx_i)[\text{erf } x - \text{erf } x_i] - \frac{k}{\sqrt{\pi}}[\exp(-x^2) - \exp(-x_i^2)]\right\}$$

$$= 0.3985\left\{3.875\left[\text{erf } x + 0.586\right] - 2.821\left[\exp(-x^2) - 0.718\right]\right\}$$

$$\Delta Y = 2.60\,\Delta Q$$

for plant K accumulation K_u and plant K concentration K_c as shown in Fig.
3.22.

Our next challenge is to link growth response of an annual crop to
applied N. Data are taken from a field study with corn by Rhoads and
Stanley (1979) as summarized by Overman and Rhoads (1991). Results
are given in Table 3.20 for dry matter and plant N accumulation. Growth
response curves are shown in Fig. 3.23 for applied N of 0 and 300 kg ha^{-1}.
Parameter values chosen for the simulation are $t_i = 18.0$ wk, $\mu = 26.0$ wk,
$\sqrt{2}\sigma = 8.0$ wk, $c = 0.2$ wk^{-1}, $k = 5$. Dimensionless time x is calculated from
Eq. [3.60], while the growth quantifier follows from

$$Q = 0.726\left\{2.00(\text{erf } x + 0.223) - 2.821\left[\exp(-x^2) - 0.9608\right]\right\} \qquad [3.72]$$

Maximum Q is calculated from Eq. [3.72] to be

Table 3.15 Model estimates of growth of coastal bermudagrass at College Station, TX for beginning time of 29.5 wk and applied N of 600 kg ha^{-1}

t wk	x	erf x	$\exp(-x^2)$	ΔQ	ΔY Mg ha^{-1}
29.5	—	—	—	—	—
31.0	1.425	0.956	0.131	0	0
32.0	1.550	0.972	0.0905	0.159	0.41
33.0	1.675	0.982	0.0605	0.387	1.01
34.0	1.800	0.989	0.0392	0.556	1.45
35.0	1.925	0.9935	0.0246	0.689	1.79
36.0	2.050	0.9962	0.0150	0.792	2.06
37.0	2.175	0.9978	0.0088	0.867	2.25
37.5	2.2375	0.9984	0.0067	0.889	2.31
38.0	2.300	0.9989	0.0050	0.906	2.36

Assume: $t_i = 31.0\,\text{wk}, \ \mu = 26.0\,\text{wk}, \ \sqrt{2}\sigma = 8.0\,\text{wk}, \ c = 0.2\,\text{wk}^{-1}, \ k = 5$

Then:
$$x = \frac{t-\mu}{\sqrt{2}\sigma} + \frac{\sqrt{2}\sigma c}{2} = \frac{t-26.0}{8.0} + 0.8 = \frac{t-19.6}{8.0}$$

$$\Delta Q = \exp\left(\sqrt{2}\sigma c x_i\right)\left\{(1-kx_i)[\text{erf }x - \text{erf }x_i] - \frac{k}{\sqrt{\pi}}[\exp(-x^2) - \exp(-x_i^2)]\right\}$$

$$= 9.777\left\{-6.125\,[\text{erf }x - 0.956] - 2.821[\exp(-x^2) - 0.131]\right\}$$

$$\Delta Y = 2.60\,\Delta Q$$

$$Q_{\max} = \lim_{x\to\infty} Q = 3.74 \qquad\qquad [3.73]$$

Since plateaus of the yield response curves are estimated to be 17.0 Mg ha^{-1} and 25.0 Mg ha^{-1} for $N = 0$ and 300 kg ha^{-1}, respectively, then yield curves are estimated from

$$N = 0:\qquad\qquad Y = AQ = (17.0/3.74)\,Q = 4.55\,Q \qquad\qquad [3.74]$$

$$N = 300\ \text{kg ha}^{-1}:\quad Y = AQ = (25.0/3.74)Q = 6.68\,Q \qquad\qquad [3.75]$$

The curves for dry matter accumulation shown in Fig. 3.23 are drawn from Eqs. [3.60], [3.72], [3.74], and [3.75]. The phase plot of plant N accumulation and dry matter accumulation is shown in Fig. 3.24. The straight lines are drawn from

Table 3.16 Model estimates of growth of coastal bermudagrass at College Station, TX for beginning time of 37.5. wk and applied N of 600 kg ha^{-1}

t wk	x	erf x	$\exp(-x^2)$	ΔQ	ΔY Mg ha^{-1}
37.5	—	—	—	—	—
39.0	2.425	0.99938	0.0028	0	0
40.0	2.550	0.99967	0.0015	0.021	0.05
41.0	2.675	0.99985	0.00078	0.023	0.06
42.0	2.800	0.999925	0.00039	0.036	0.09
43.0	2.925	0.999965	0.00019	0.041	0.11
44.0	3.050	0.999988	0.000091	0.043	0.11
45.0	3.175	1.00000	0.000042	0.043	0.11
45.5	3.2375	1.00000	0.000028	0.045	0.12
46.0	3.300	1.00000	0.000019	0.046	0.12

Assume: $t_i = 39.0\,\text{wk}, \ \mu = 26.0\,\text{wk}, \ \sqrt{2}\sigma = 8.0\,\text{wk}, \ c = 0.2\,\text{wk}^{-1}, \ k = 5$

Then: $$x = \frac{t-\mu}{\sqrt{2}\sigma} + \frac{\sqrt{2}\sigma c}{2} = \frac{t-26.0}{8.0} + 0.8 = \frac{t-19.6}{8.0}$$

$$\Delta Q = \exp\left(\sqrt{2}\sigma c x_i\right)\left\{(1 - kx_i)[\text{erf } x - \text{erf } x_i] - \frac{k}{\sqrt{\pi}}[\exp(-x^2) - \exp(-x_i^2)]\right\}$$

$$= 48.42\left\{-11.125\,[\text{erf } x - 0.99952] - 2.821[\exp(-x^2) - 0.0028]\right\}$$

$$\Delta Y = 2.60\,\Delta Q$$

$N = 0:$ $\qquad\qquad \dfrac{Y}{N_u} = 0.0468 + 0.00745\,Y \qquad r = 0.9794 \qquad$ [3.76]

$N = 300 \text{ kg ha}^{-1}:$ $\ \dfrac{Y}{N_u} = 0.0348 + 0.00349\,Y \qquad r = 0.9721 \qquad$ [3.77]

As a consequence the curves in Fig. 3.24 are drawn from

$N = 0:$ $\qquad\qquad N_u = \dfrac{134\,Y}{6.3 + Y}$ $\qquad\qquad\qquad\qquad\qquad$ [3.78]

$N = 300 \text{ kg ha}^{-1}: N_u = \dfrac{287\,Y}{10.0 + Y}$ $\qquad\qquad\qquad\qquad$ [3.79]

Estimated Y values are then entered into Eqs [3.78] and [3.79] to calculate N_u to draw the curves for plant N accumulation in Fig. 3.23. Then plant N concentrations are estimated from

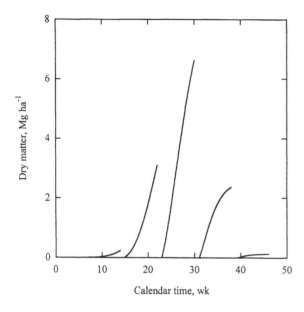

Figure 3.17 Simulated growth curves for coastal bermudagrass for study of Villanueva (1974). Values are taken from Tables 3.12 through 3.16.

Table 3.17 Normalized distribution of the growth function for coastal bermudagrass with harvest interval of 8 wk at College Station, TX

t wk	ΔY Mg ha^{-1}	Y_n Mg ha^{-1}	F	erf$^{-1}(2F - 1)$
5.5		0	0	
	0.19			
13.5		0.19	0.016	−1.51
	2.74			
21.5		2.93	0.252	−0.472
	6.27			
29.5		9.20	0.791	0.573
	2.31			
37.5		11.51	0.990	1.63
	0.12			
45.5		11.63	1	

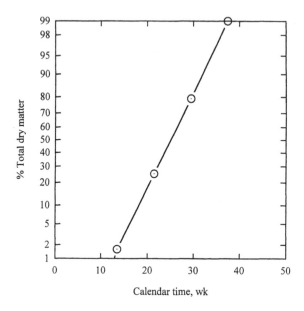

Figure 3.18 Probability plot of dry matter distribution for coastal bermudagrass from study of Villanueva (1974). Values are taken from Table 3.17.

Table 3.18 Growth response of corn at Florence, SC

t wk	Y Mg ha^{-1}	N_u kg ha^{-1}	N_c g kg^{-1}	P_u kg ha^{-1}	P_c g kg^{-1}	K_u kg ha^{-1}	K_c g kg^{-1}
15.0	—	—	—	—	—	—	—
17.8	0.042	1	—	—	—	—	—
19.5	0.406	13	32.0	1	2.5	18	44.2
21.0	1.83	60	32.8	6	3.3	118	64.5
22.0	4.21	110	26.1	13	3.1	219	52.1
24.0	10.2	171	16.8	22	2.2	298	29.2
26.8	18.8	198	10.5	32	1.7	305	16.2
28.8	21.3	204	9.6	34	1.6	267	12.5
29.5	22.5	207	9.2	34	1.5	293	13.0

Data adapted from Karlen et al. (1987).

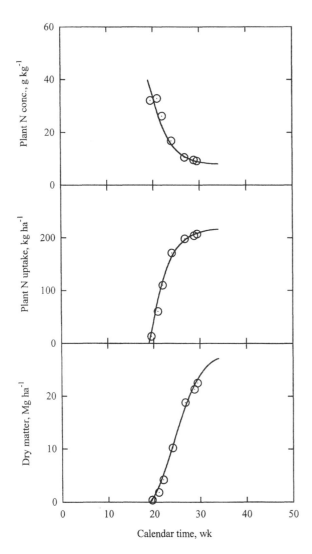

Figure 3.19 Dry matter and plant N accumulation with time by corn. Data from Karlen et al. (1987). Curves drawn from Eqs. [3.60] through [3.62], [3.66], and [3.67].

Table 3.19 Model simulation for corn at Florence, SC

t wk	x	erf x	exp $(-x^2)$	Q	Y Mg ha^{-1}	N_u kg ha^{-1}	N_c g kg^{-1}
19	−0.075	−0.085	0.9944	0	0	0	39.7
20	0.050	0.056	0.9975	0.164	1.20	40	33.8
21	0.175	0.201	0.970	0.410	3.01	83	27.5
22	0.300	0.329	0.914	0.706	5.18	117	22.5
23	0.425	0.452	0.835	1.054	7.74	143	18.6
24	0.550	0.563	0.739	1.429	10.49	164	15.6
25	0.675	0.660	0.634	1.810	13.29	179	13.4
26	0.800	0.742	0.527	2.178	15.99	189	11.8
27	0.925	0.810	0.425	2.516	18.47	197	10.7
28	1.050	0.862	0.332	2.812	20.64	203	9.8
29	1.175	0.903	0.251	3.065	22.50	207	9.2
30	1.300	0.934	0.185	3.268	23.99	210	8.8
32	1.550	0.972	0.090	3.552	26.07	214	8.2
34	1.800	0.989	0.039	3.700	27.16	216	8.0
		1	0	3.811	27.97	217	7.8

Assume: $t_i = 19.0$ wk, $\mu = 26.0$ wk, $\sqrt{2}\sigma = 8.0$ wk, $c = 0.2$ wk^{-1}, $k = 5$

Then: $$x = \frac{t - \mu}{\sqrt{2}\sigma} + \frac{\sqrt{2}\sigma c}{2} = \frac{t - 26.0}{8.0} + 0.8 = \frac{t - 19.6}{8.0}$$

$$\Delta Q = \exp\left(\sqrt{2}\sigma c x_i\right)\left\{(1 - kx_i)[\text{erf } x - \text{erf } x_i] - \frac{k}{\sqrt{\pi}}[\exp(-x^2) - \exp(-x_i^2)]\right\}$$

$$= 0.887\left\{1.375\,[\text{erf } x + 0.085] - 2.821[\exp(-x^2) - 0.9944]\right\}$$

$$\Delta Y = \left(\frac{24.0}{3.27}\right)\Delta Q = 7.34\,\Delta Q$$

$N = 0:$ $$N_c = \frac{134}{6.3 + Y}$$ [3.80]

$N = 300$ kg ha^{-1} : $$N_c = \frac{287}{10.0 + Y}$$ [3.81]

as shown in Fig. 3.23. It can be seen from Figs. 3.23 and 3.24 that the model provides excellent simulation of the growth curves. The simulation procedure is summarized in Table 3.21 for $N = 0$. Now it remains to demonstrate the general response of dry matter and plant N uptake to applied N. From the maximum values for yield and plant N uptake, Y and N_u, respectively,

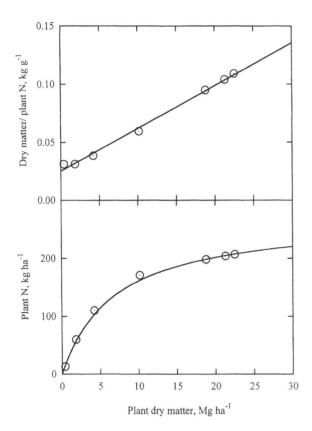

Figure 3.20 Dependence of plant N accumulation on plant dry matter accumulation by corn from study of Karlen et al. (1987). Curve drawn from Eq. [3.66]; line from Eq. [3.65].

listed in Table 3.20, we can plot the values shown in Fig. 3.25 where the maximum plant N concentration N_{cm} is calculated from

$$N_{c\infty} = \frac{N_{u\infty}}{Y_{\infty}}$$ [3.82]

Since Overman et al. (1994) have shown previously that total yield and plant N uptake can be described by the logistic equation, we use this approach to obtain the estimation equations

$$Y_{\infty} = \frac{25.0}{1 + \exp(-0.75 - 0.0100N)}$$ [3.83]

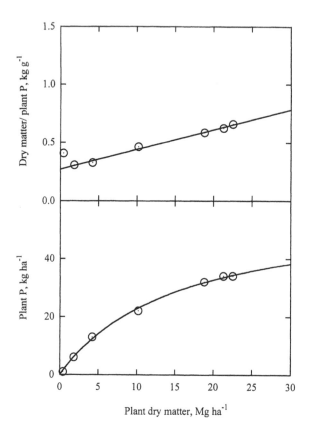

Figure 3.21   Dependence of plant P accumulation on plant dry matter accumulation by corn from study of Karlen et al. (1987). Curve drawn from Eq. [3.69]; line from Eq. [3.68].

$$N_{u\infty} = \frac{230}{1 + \exp(0.10 - 0.0100N)} \qquad [3.84]$$

$$N_{c\infty} = 9.20 \left[\frac{1 + \exp(-0.75 - 0.0100N)}{1 + \exp(0.10 - 0.0100N)} \right] \qquad [3.85]$$

The curves in Fig. 3.25 are drawn from Eqs. [3.83] through [3.85]. The phase relationship Y_∞ vs. $N_{u\infty}$, is shown in Fig. 3.26. The curve is drawn from the hyperbolic equation

$$Y_\infty = \frac{43.7\, N_{u\infty}}{172 + N_{u\infty}} \qquad [3.86]$$

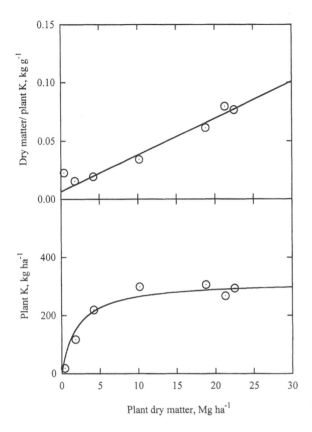

Figure 3.22 Dependence of plant K accumulation on plant dry matter accumulation by corn from study of Karlen et al. (1987). Curve drawn from Eq. [3.71]; line from Eq. [3.70].

as discussed by Overman et al. (1994). It follows from Eq. [3.86] that maximum plant N concentration can be estimated as shown in Fig. 3.26 from

$$N_{c\infty} = 3.94 + 0.0229\, N_{u\infty} \qquad [3.87]$$

Refinement of the phase relationship between plant N and dry matter accumulation with time is now required. A summary of the hyperbolic parameters is given in Table 3.22. Since there is no clear trend in K_y with applied N, we will choose the average value of $K_y = 11.0$ Mg ha^{-1} and then standardize N_u to obtain estimates of N_{um} for each applied N. The phase equation is now written as

Table 3.20 Dry matter and plant nutrient accumulation by corn at Quincy, FL

N, kg ha^{-1}	t, wk	18.7	20.4	22.7	25.7	max
0	Y, Mg ha^{-1}	0.4	3.1	4.9	10.2	17.0
	N_c, g kg^{-1}	22.0	12.8	12.2	8.2	—
	N_u, kg ha^{-1}	9	39	60	84	98
	Y/N_u, kg g^{-1}	0.044	0.079	0.082	0.121	—
60	Y, Mg ha^{-1}	0.72	2.6	5.1	11.1	19.0
	N_c, g kg^{-1}	22.0	14.6	13.7	11.0	—
	N_u, kg ha^{-1}	16	39	70	122	150
	Y/N_u, kg g^{-1}	0.045	0.067	0.073	0.091	—
120	Y, Mg ha^{-1}	0.78	2.7	6.6	12.7	21.5
	N_c, g kg^{-1}	23.9	17.8	14.1	11.4	—
	N_u, kg ha^{-1}	19	48	93	144	175
	Y/N_u, kg g^{-1}	0.041	0.056	0.071	0.088	—
180	N_c, Mg ha^{-1}	1.0	3.2	7.0	13.4	23.0
	N_c, g kg^{-1}	23.2	18.6	13.3	12.3	—
	N_u, kg ha^{-1}	24	59	93	165	200
	Y/N_u, kg g^{-1}	0.042	0.054	0.075	0.081	—
300	Y, Mg ha^{-1}	1.0	3.8	8.0	14.9	25.0
	N_c, g kg^{-1}	27.6	22.4	14.3	11.9	—
	N_u, kg ha^{-1}	26	86	114	177	205
	Y/N_u, kg g^{-1}	0.038	0.044	0.070	0.084	—

Data from Overman and Rhoads (1991). Numbers in last column represent estimated maximum values.

$$N_u = \frac{N_{um}\, Y}{11.0 + Y} \qquad [3.88]$$

which can be rearranged to obtain

$$N_{um} = N_u \frac{11.0 + Y}{Y} \qquad [3.89]$$

Estimates of N_{um} according to Eq. [3.89] are listed in Table 3.23. Average values of N_{um} vs. applied N are shown in Fig. 3.27, where the curve is drawn from

$$N_{um} = \frac{300}{1 + \exp(-0.75 - 0.0100N)} \qquad [3.90]$$

Equations [3.88] and [3.90] can now be coupled to link plant N to dry matter for each applied N. We have now established that the linkage between plant N and dry matter accumulation with time is described by the hyperbolic relationship

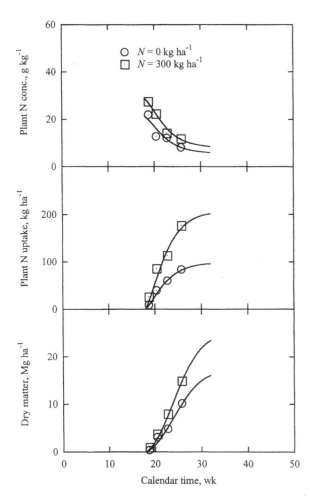

Figure 3.23 Dry matter and plant N accumulation with time by corn. Data from Rhoads and Stanley (1979) as reported by Overman and Rhoads (1991). Curves drawn from Eqs. [3.60], [3.72] through [3.81].

$$N_u = \frac{N_{um} Y}{K_y + Y}$$
[3.91]

It has also been shown that seasonal total dry matter and plant N uptake in response to applied N follows another hyperbolic relationship:

$$Y = \frac{Y_m N_u}{K_n + N_u}$$
[3.92]

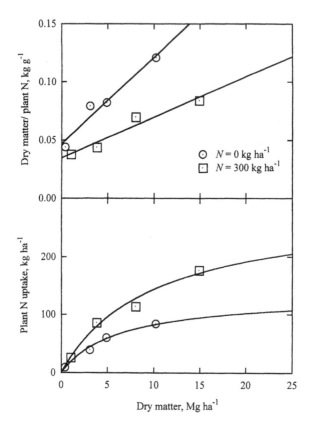

Figure 3.24 Dependence of plant N accumulation on plant dry matter accumulation by corn from Overman and Rhoads (1991). Curves drawn from Eqs. [3.78] and [3.79]; lines from Eqs. [3.76] and [3.77].

Note that the phase relationship given by Eq. [3.91] is implicit in time, while the phase relationship given by Eq. [3.92] is implicit in applied N. The question naturally arises as to how these equations couple to control the maximum estimates of Y and N_u listed in Table 3.20. Equations [3.91] and [3.92] can be solved simultaneously to obtain

$$Y' = \frac{Y_m N_{um} - K_y K_n}{N_{um} + K_n}$$ [3.93]

$$N_u' = \frac{Y_m N_{um} - K_y K_n}{Y_m + K_y}$$ [3.94]

Table 3.21 Model estimates for corn at Quincy, FL for no applied N

t wk	x	erf x	$\exp(-x^2)$	Q	Y Mg ha^{-1}	N_u kg ha^{-1}	N_c g kg^{-1}
18	−0.200	−0.223	0.9608	0	0	0	21.3
19	−0.075	−0.085	0.9944	0.132	0.60	12	19.4
20	0.050	0.056	0.9975	0.330	1.50	26	17.2
21	0.175	0.195	0.970	0.588	2.67	40	14.9
22	0.300	0.329	0.914	0.897	4.08	53	12.9
23	0.425	0.453	0.835	1.24	5.64	63	11.2
24	0.550	0.563	0.739	1.60	7.27	72	9.9
25	0.675	0.660	0.634	1.95	8.86	78	8.8
26	0.800	0.742	0.527	2.29	10.4	83	8.0
27	0.925	0.810	0.425	2.60	11.8	87	7.4
28	1.050	0.862	0.332	2.86	13.0	90	6.9
29	1.175	0.904	0.251	3.09	14.0	92	6.6
30	1.300	0.934	0.185	3.27	14.9	94	6.3
31	1.425	0.956	0.131	3.41	15.5	95	6.1
32	1.550	0.972	0.090	3.52	16.0	96	6.0
∞	∞	1	0	3.74	17.0	98	5.8

Assume: $t_i = 18.0\,\text{wk}$, $\mu = 26.0\,\text{wk}$, $\sqrt{2}\sigma = 8.0\,\text{wk}$, $c = 0.2\,\text{wk}^{-1}$, $k = 5$

Then: $$x = \frac{t-\mu}{\sqrt{2}\sigma} + \frac{\sqrt{2}\sigma c}{2} = \frac{t-26.0}{8.0} + 0.8 = \frac{t-19.6}{8.0}$$

$$\Delta Q = \exp\left(\sqrt{2}\sigma c x_i\right)\left\{(1 - kx_i)[\text{erf } x - \text{erf } x_i] - \frac{k}{\sqrt{\pi}}[\exp(-x^2) - \exp(-x_i^2)]\right\}$$

$$= 0.726\left\{2.00\,[\text{erf } x + 0.223] - 2.821[\exp(-x^2) - 0.9608]\right\}$$

$$\Delta Y = \left(\frac{17.0}{3.74}\right)\Delta Q = 4.55\ \Delta Q$$

A summary of parameters and estimates of maxima is given in Table 3.22, where the maxima are calculated from Eqs. [3.93] and [3.94]. A graphical alternative for obtaining Y' and N_u' is illustrated in Fig. 3.28, where the curves are drawn from Eqs. [3.88] and [3.90]. The points of intersection of the curves represent the values for Y' and N_u' for each applied N. It should now become apparent that the two hyperbolic relationships serve to make the plant system self-limiting.

The growth model also applies to perennial ryegrass. Data from a study by Green and Corrall as reported in Whitehead (1995) are given in Table 3.24 and shown in Fig. 3.29. Dry matter accumulation is estimated from the

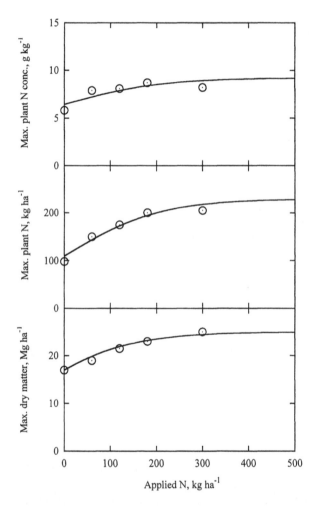

Figure 3.25 Dependence of maximum dry matter, plant N uptake, and plant N concentration on applied N for corn study from Overman and Rhoads (1991). Curves drawn from Eqs. [3.83] through [3.85].

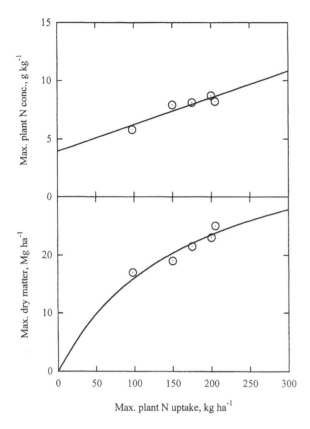

Figure 3.26 Dependence of maximum dry matter on maximum plant N uptake for corn study from Overman and Rhoads (1991). Curve drawn from Eq. [3.86]; line from Eq. [3.87].

Table 3.22 Summary of hyperbolic parameters and maximum accumulation of dry matter and plant N for corn at Quincy, FL

N kg ha^{-1}	Y_m Mg ha^{-1}	K_n kg ha^{-1}	N_{um} kg ha^{-1}	K_y Mg ha^{-1}	Y Mg ha^{-1}	N_u kg ha^{-1}
—	43.7	172	—	—	—	—
0	—	—	134	6.3	15.5	95
60	—	—	253	12.6	20.9	158
120	—	—	267	11.4	22.1	176
180	—	—	321	14.1	23.6	201
300	—	—	287	10.0	23.6	202
Avg.	—	—	—	11.0	—	—

Table 3.23 Estimated dry matter and plant nutrient accumulation by corn at Quincy, FL

N, kg ha^{-1}	t, wk	18.7	20.4	22.7	25.7	$N_{um\ avg}$
0	Y, Mg ha^{-1}	0.4	3.1	4.9	10.2	—
	N_u, kg ha^{-1}	9	39	60	84	—
	N_{um}, kg ha^{-1}	260	180	200	180	200
60	Y, Mg ha^{-1}	0.72	2.6	5.1	11.1	—
	N_u, kg ha^{-1}	16	39	70	122	—
	N_{um}, kg ha^{-1}	260	200	220	240	230
120	Y, Mg ha^{-1}	0.78	2.7	6.6	12.7	—
	N_u, kg ha^{-1}	19	48	93	144	—
	N_{um}, kg ha^{-1}	290	240	250	270	260
180	Y, Mg ha^{-1}	1.0	3.2	7.0	13.4	—
	N_u, kg ha^{-1}	24	59	93	165	—
	N_{um}, kg ha^{-1}	290	260	240	300	270
300	Y, Mg ha^{-1}	1.0	3.8	8.0	14.9	—
	N_u, kg ha^{-1}	26	86	114	177	—
	N_{um}, kg ha^{-1}	310	330	270	310	300

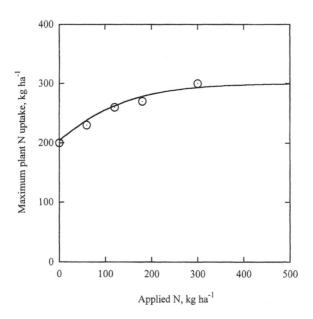

Figure 3.27 Dependence of maximum plant N uptake on applied N for corn study from Overman and Rhoads (1991). Curve drawn from Eq. [3.90].

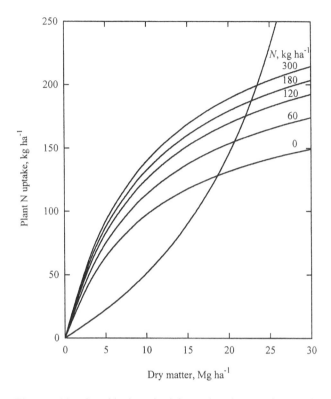

Figure 3.28 Graphical method for estimating maximum plant N uptake and maximum dry matter as related to applied N. Curves are drawn from Eqs. [3.86], [3.88], and [3.90].

Table 3.24 Dry matter and plant N accumulation by perennial ryegrass at Hurley, England

t wk	Y Mg ha^{-1}	N_u kg ha^{-1}	N_c g kg^{-1}	Y/N_u kg g^{-1}
14.4	1.0	48	48	0.021
15.9	2.3	83	36	0.028
17.3	4.0	110	28	0.036
18.7	6.0	120	20	0.050
20.1	7.9	130	17	0.059
21.6	9.2	120	13	0.077
23.1	9.8	120	12	0.082
24.6	9.9	110	11	0.090

Data of Green and Corrall (unpublished) as reported by Whitehead (1995).

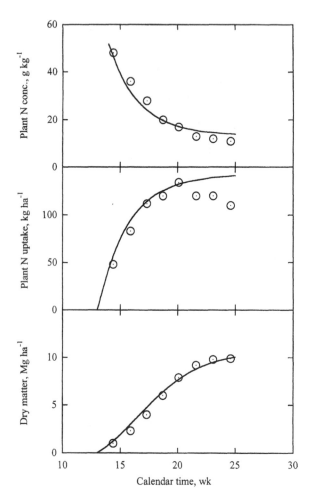

Figure 3.29 Dry matter and plant N accumulation by perennial ryegrass at Hurley, England. Curves drawn from Eqs. [3.95] through [3.97], [3.99], and [3.100].

model with the following parameters: $t_i = 13.0$ wk, $\mu = 26.0$ wk, $\sqrt{2}\mu = 8.0$ wk, $c = 0.5$ wk^{-1}, $k = 5$. Dimensionless time is defined by Eq. [3.41] and becomes

$$x = \frac{t - 10.0}{8.0} \qquad\qquad [3.95]$$

and the growth quantifier, defined by Eq. [3.40], becomes

$$Q = 4.48 \left\{ -0.875(\text{erf } x - 0.404) - 2.821\left[\exp(-x^2) - 0.869\right] \right\} \qquad [3.96]$$

Model simulation is given in Table 3.25. Dry matter accumulation is calculated from

$$Y = (10.5/8.65)\, Q = 1.21\, Q \qquad [3.97]$$

The phase relation between plant N accumulation and dry matter accumulation is shown in Fig. 3.30, where the line and curve are drawn from

Table 3.25 Model estimates of growth of perennial ryegrass at Hurley, England

t wk	x	erf x	exp $(-x^2)$	ΔQ	Y Mg ha^{-1}	N_u kg ha^{-1}	N_c g kg^{-1}
13	0.375	0.404	0.869	0	0	0	68.1
14	0.500	0.520	0.779	0.68	0.82	42.4	51.8
15	0.625	0.623	0.677	1.57	1.90	74.7	39.3
16	0.750	0.711	0.570	2.58	3.12	96.5	30.9
17	0.875	0.784	0.465	3.62	4.38	111	25.4
18	1.000	0.843	0.368	4.61	5.58	121	21.6
19	1.125	0.888	0.282	5.52	6.68	127	19.1
20	1.250	0.923	0.210	6.29	7.61	132	17.3
21	1.375	0.948	0.151	6.94	8.40	135	16.1
22	1.500	0.966	0.105	7.45	9.01	136	15.2
23	1.625	0.9784	0.071	7.83	9.49	139	14.7
24	1.750	0.9867	0.0468	8.11	9.81	140	14.3
25	1.875	0.9920	0.0297	8.30	10.04	141	14.0
26	2.000	0.9953	0.0183	8.43	10.20	142	13.9
		1	0	8.65	10.47	142	13.5

Assume: $t_i = 13.0\,\text{wk}$, $\mu = 26.0\,\text{wk}$, $\sqrt{2}\sigma = 8.0\,\text{wk}$, $c = 0.5\,\text{wk}^{-1}$, $k = 5$

Then: $\quad x = \dfrac{t - \mu}{\sqrt{2}\sigma} + \dfrac{\sqrt{2}\sigma c}{2} = \dfrac{t - 26.0}{8.0} + 2.0 = \dfrac{t - 10.0}{8.0}$

$$\Delta Q = \exp\left(\sqrt{2}\sigma c x_i\right)\left\{(1 - kx_i)[\text{erf } x - \text{erf } x_i] - \dfrac{k}{\sqrt{\pi}}[\exp(-x^2) - \exp(-x_i^2)]\right\}$$

$$= 4.48\left\{-0.875\,[\text{erf } x - 0.404] - 2.821[\exp(-x^2) - 0.869]\right\}$$

$$\Delta Y = \left(\dfrac{10.5}{8.65}\right)\Delta Q = 1.21\,\Delta Q$$

$$N_u = \dfrac{177\,Y}{2.60 + Y}$$

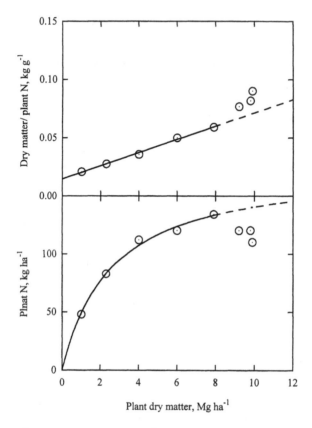

Figure 3.30 Dependence of plant N accumulation on plant dry matter accumulation by perennial ryegrass from report of Whitehead (1995). Curve drawn from Eq. [3.99]; line from Eq. [3.98].

$$\frac{Y}{N_u} = 0.0147 + 0.00566\,Y \qquad r = 0.9976 \tag{3.98}$$

$$N_u = \frac{N_{um}\,Y}{K_y + Y} = \frac{177\,Y}{2.60 + Y} \tag{3.99}$$

where the last three data points have been omitted from regression analysis. Plant N concentration is estimated from

$$N_c = \frac{N_{um}}{K_y + Y} = \frac{177}{2.60 + Y} \tag{3.100}$$

The curves in Fig. 3.29 are drawn from Eqs. [3.97], [3.99], and [3.100]. Apparently after 8 weeks of regrowth the plants begin to lose nitrogen.

EXERCISES

3.1 Prins et al. (1980) measured dry matter accumulation with time by grass (unspecified species) in the Netherlands. Data are given in the table below. Applied $N = 640$ kg ha^{-1}. Maximum dry matter is estimated as 16.50 Mg ha^{-1}.

Dependence of dry matter accumulation on time for grass in the Netherlands

t wk	ΔY Mg ha^{-1}	Y Mg ha^{-1}	F	z
10.1		0	0	
	1.72			
16.7		1.72	0.104	-0.89
	2.45			
20.1		4.17	0.253	-0.47
	2.32			
24.1		6.49	0.393	-0.192
	2.20			
27.1		8.69	0.527	0.048
	1.88			
29.7		10.57	0.641	0.255
	1.84			
32.4		12.41	0.752	0.482
	1.99			
36.1		14.40	0.873	0.807
	1.67			
43.4		16.07	0.974	1.375
∞		16.50		

Data adapted from Prins et al. (1980).

a. Plot cumulative dry matter Y vs. time t on linear graph paper.
b. Plot F vs. t on probability paper.
c. Perform linear regression on z vs. t and estimate parameters μ and σ.
d. Assume model parameters: $\mu = 26.5$ wk, $\sqrt{2}\sigma = 12.0$ wk, $c = 0.2$ wk^{-1}, $k = 5$ to calculate the growth quantifier and cumulative dry matter with time. Use $A = 1.50$ Mg ha^{-1} for dry matter simulation.

e. Plot (d) on (a).

f. Discuss the results.

3.2 Deinum and Sibma (1980) measured dry matter and plant N accumulation with time by grass (unspecified species) in the Netherlands. Data
are given in the table below. Applied N = 200 kg ha^{-1}. Maximum dry
matter is estimated as 8.00 Mg ha^{-1}.

Dependence of dry matter and plant N on time for grass grown in
the Netherlands

t wk	Y Mg ha^{-1}	N_u kg ha^{-1}	N_c g kg^{-1}	Y/N_u kg g^{-1}
19.0	—	—	—	—
21.4	0.58	32	55.2	0.0181
22.4	1.91	94	49.2	0.0203
23.6	3.58	144	40.2	0.0249
24.6	5.28	168	31.8	0.0314
25.6	6.20	177	28.5	0.0350
27.4	6.87	168	24.5	0.0409
—	8.00	—	—	—

Yield and plant N data adapted from Deinum and Sibma (1980).

a. Plot cumulative dry matter Y, plant N uptake N_u, and plant N
concentration N_c vs. time t on linear graph paper.

b. Assume model parameters $t_i = 21$ wk, $\mu = 26.0$ wk, $\sqrt{2}\sigma = 8.0$ wk,
$c = 0.5$ wk^{-1}, $k = 5$ to calculate the growth quantifier and cumulative dry matter with time.

c. Plot (b) on (a).

d. Plot N_u and Y/N_u vs. Y on linear graph paper.

e. Perform linear regression on Y/N_u vs. Y.

f. Calculate parameters N_{um} and K_y from (e).

g. Use the parameters from (f) to write the phase equation.

h. Estimate values for N_u from estimated Y values of the model.

i. Estimate the corresponding values of N_c for each time.

j. Plot N_u and N_c vs. t on (d).

k. Discuss the results.

3.3 Blue (1987) studied accumulation of dry matter and plant N with time
by pensacola bahiagrass on the Entisol Astatula sand (hyperthermic,
uncoated Typic Quartzipsamments) and the Spodosol Myakka fine
sand (sandy, siliceous, hyperthermic Aeric Haplaquods) near

Gainesville, FL. Data for 1979 with applied N of 300 kg ha^{-1} and average harvest interval of 6.6 wk is given in the table below, taken from Overman and Blue (1991).

a. Plot dry matter Y, plant N uptake N_u, and plant N concentration N_c vs. time t for both soils on linear graph paper.
b. Use the expanded growth model to calculate Y vs. t with the parameters $\mu = 26.0$ wk, $\sqrt{2}\sigma = 11.0$ wk, $c = 0.2$ wk^{-1}, $k = 5$. Use A(Entisol) $= 1.25$ Mg ha^{-1} and A(Spodosol) $= 2.35$ Mg ha^{-1}.
c. Plot results of (b) on (a).
d. Plot F_n vs. F_y on linear graph paper. Calculate and plot the regression line.
e. Calculate N_u vs. t and plot the results on (a).
f. Discuss the results. How well does the model fit the data?

Dry matter and plant N accumulation with time for bahiagrass grown with N = 300 kg ha^{-1} on two soils near Gainesville, FL

Soil	t wk	ΔY Mg ha^{-1}	Y Mg ha^{-1}	F_y	ΔN_u kg ha^{-1}	N_u kg ha^{-1}	F_n	N_c g kg^{-1}
Entisol	—		0	0		0	0	
		1.99			39.4			19.8
	19.3		1.99	0.187		39.4	0.214	
		1.58			22.0			13.9
	26.0		3.57	0.366		61.4	0.334	
		3.41			64.0			18.8
	31.7		6.98	0.657		125.4	0.682	
		2.75			43.2			15.7
	39.1		9.73	0.915		168.6	0.916	
		0.90			15.4			17.1
	—		10.63	1		184.0	1	
Spodosol	—		0	0		0	0	
		4.79			67.5			14.1
	19.3		4.79	0.268		67.5	0.285	
		2.92			34.0			11.6
	26.0		7.71	0.431		101.5	0.428	
		6.18			84.0			13.6
	31.7		13.89	0.777		185.5	0.783	
		3.08			39.7			12.9
	39.1		16.97	0.950		225.2	0.950	
		0.90			11.8			13.1
	—		17.87	1		237.0	1	

Data for 1979 from Overman and Blue (1991).

3.4 Sayre (1948) measured accumulation of dry matter and plant nutrients with time (days after planting) by corn at Wooster, OH. Data are listed in the table below.

   a. Assume time of planting was 20 wk after Jan. 1. Calculate t, wk for each sampling.
   b. Assume the parameters $t_i = 26$ wk, $\mu = 28$ wk, $\sqrt{2}\sigma = 8$ wk, $c = 0.2$ wk^{-1}, $k = 5$. Show that Q is given by $Q = 2.41\ \{-1.75(\text{erf}\,x - 0.563) - 2.821[\exp(-x^2) - 0.739]\}$. Calculate x and Q for each sampling date.
   c. Assume the yield plateau is 14.0 Mg ha^{-1}. Show that yield Y can be estimated from $Y = 4.40\,Q$. Calculate Y for $t \geq 26$ wk.
   d. Show that dry matter and plant N accumulation are related by $Y/N_u = 0.0355 + 0.00355\,Y$ and that plant N uptake N_u is given by $N_u = 282Y/(10.0 + Y)$. Calculate plant N uptake for $t \geq 26$ wk. Calculate plant N concentration N_c for $t \geq 26$ wk.
   e. Plot measured and estimated values for Y, N_u and N_c vs. t on linear graph paper.
   f. Obtain the equation for plant P uptake P_u, $Y/P_u = a + b\,Y$, by linear regression. Calculate plant P uptake for $t \geq 26$ wk. Calculate plant P concentration, P_c, for $t \geq 26$ wk.
   g. Plot measured and estimated values for Y, P_u and P_c vs. t on linear graph paper.
   h. Obtain the equation for plant K uptake K_u, $Y/K_u = a + b\,Y$, by linear regression. Calculate plant K uptake for $t \geq 26$ wk. Calculate plant K concentration K_c for $t \geq 26$ wk.
   i. Plot measured and estimated values for Y, K_u and K_c vs. t on linear graph paper.
   j. Discuss the results.

Data for corn study at Wooster, OH

t d	Y Mg ha^{-1}	N_u kg ha^{-1}	N_c g kg^{-1}	P_u kg ha^{-1}	P_c g kg^{-1}	K_u kg ha^{-1}	K_c g kg^{-1}
0	—	—	—	—	—	—	—
28	0.132	3.9	29.7	0.3	2.55	5.5	41.5
31	0.211	5.8	27.7	0.6	2.65	9.2	43.6
34	0.237	7.4	31.1	0.8	3.31	10.3	43.4
37	0.501	14.6	29.0	1.5	2.91	17.1	34.2
40	0.633	18.7	29.6	1.9	3.01	26.9	42.5
43	0.870	24.5	28.2	2.7	3.09	36.2	41.6
46	1.24	31.1	25.1	3.7	2.98	46.7	37.7
49	1.66	39.3	23.7	4.8	2.90	57.5	34.6

52	2.19	47.7	21.8	6.2	2.81	68.3	31.2
55	2.74	57.1	20.8	7.7	2.82	79.4	29.0
58	3.35	66.0	19.7	9.9	2.94	89.4	26.7
61	3.96	74.6	18.8	11.1	2.80	98.1	24.8
64	4.54	82.5	18.2	12.5	2.76	104	22.9
67	5.30	93.6	17.7	15.1	2.85	111	21.0
70	6.25	107	17.2	17.1	2.74	118	18.9
73	6.88	110	15.9	18.9	2.75	122	17.7
76	7.36	111	15.0	19.6	2.66	124	16.9
79	8.04	120	14.9	21.2	2.63	126	15.7
82	8.62	130	15.1	23.2	2.69	127	14.7
85	9.31	135	14.5	24.1	2.59	127	13.6
88	9.76	141	14.4	25.1	2.57	127	13.0
91	10.23	149	14.6	25.5	2.50	127	12.3
94	10.76	156	14.5	26.7	2.48	127	11.8
97	11.24	159	14.1	28.2	2.51	124	11.0
100	11.66	159	13.6	29.7	2.55	119	10.2
103	12.00	158	13.2	30.9	2.58	114	9.5
106	12.40	156	12.6	31.5	2.54	112	9.0
109	12.82	157	12.2	31.9	2.49	110	8.6
112	13.27	158	11.9	32.7	2.46	110	8.3
115	13.66	160	11.7	34.5	2.53	110	8.1
118	13.87	161	11.6	33.8	2.44	110	7.9

Adapted from Sayre (1948).

3.5 Bar-Yosef and Kafkafi (1972) measured growth of Newe Ya'ar hybrid corn at Bet Dagan, Israel in response to applied N and P. The soil was montmorillonitic clay, deep alluvial grumusol. Corn was planted on July 18, 1968 (28.6 wk after Jan. 1) and later thinned to a population of 50,000 plants ha^{-1}. Data are listed for N = 200 kg ha^{-1} and P = 80 kg ha^{-1}.

a. Assume the parameters $t_i = 32.5$ wk, $\mu = 32$ wk, $\sqrt{2}\sigma = 8$ wk, $c = 0.2$ wk^{-1}, $k = 5$. Show that Q is given by $Q = 3.975\{-3.312(erf x -0.7775) - 2.821[exp(-x^2) - 0.4752]\}$. Calculate x and Q for each sampling date.

b. Assume the yield is 14.0 Mg ha^{-1} at $t = 40$ wk. Show that yield Y can be estimated from $Y = 6.66\,Q$. Calculate Y for each sampling date.

c. Show that dry matter and plant N accumulation are related by $Y/N_u = 0.0200 + 0.00381\,Y$ and that plant N uptake N_u is given by $N_u = 262Y/(5.2 + Y)$. Calculate plant N uptake for each sampling date. Calculate plant N concentration N_c for each sampling date.

    d. Plot measured and estimated values for Y, N_u and N_c vs. t on linear graph paper.

    e. Discuss the results.

Data for corn study at Bet Dagan, Israel for N = 200 kg ha^{-1} and P = 80 kg ha^{-1}.

t wk	Y Mg ha^{-1}	N_u kg ha^{-1}	N_c g kg^{-1}	P_u kg ha^{-1}	P_c g kg^{-1}
28.6	—	—	—	—	—
30.3	0.01	0.5	45	0.03	2.7
31.3	0.06	3	44	0.3	4.5
32.5	0.40	16	39	1.7	4.3
33.3	1.52	58	38	5.8	3.8
34.3	3.46	110	32	11	3.2
35.3	4.90	132	27	12	2.5
36.3	8.42	200	24	18	2.1
37.3	9.95	210	21	18	1.8
38.3	11.44	170	15	21	1.8
39.3	14.21	180	13	27	1.9
40.5	14.54	190	13	26	1.8
41.5	15.50	190	12	28	1.8

Data adapted from Bar-Yosef and Kafkafi (1972).

3.6 Overman (1999) analyzed accumulation of dry matter and plant nutrients by tobacco (*Nicotiana tabacum*), a warm-season, broad-leafed annual. Data from the field study in North Carolina are given in the table below.

    a. Calculate plant nutrient uptake for N, P, K, Ca, and Mg.

    b. Plot dry matter Y, plant N uptake N_u, and plant N concentration N_c vs. time on linear graph paper.

    c. Calculate dimensionless time x and growth quantifier Q for the parameters $t_i = 22$ wk, $\mu = 28$ wk, $\sqrt{2}\sigma = 8$ wk, $c = 0.2$ wk^{-1}, $k = 5$.

    d. Calculate dry matter accumulation from $Y = 1.444Q$.

    e. Calculate plant N accumulation from $N_u = 80.0\,Y/(1.21 + Y)$.

    f. Calculate plant N concentration from $N_c = 80.0/(1.21 + Y)$.

    g. Plot Y, N_u and N_c vs. t on (b).

    h. Plot N_u and N_c vs. Y on linear graph paper for data and model.

    i. Calculate plant P uptake from $P_u = 18.7\,Y/(5.26 + Y)$.

    j. Plot P_u and P_c vs. Y on linear graph paper for data and model.

k. Calculate plant K uptake from $K_u = 235\, Y/(3.29 + Y)$.
l. Plot K_u and K_c vs. Y on linear graph paper for data and model.
m. Calculate plant Ca uptake from $Ca_u = 61.7\, Y/(3.80 + Y)$.
n. Plot Ca_u and Ca_c vs. Y on linear graph paper for data and model.
o. Calculate plant Mg uptake from $Mg_u = 21.9\, Y/(4.30 + Y)$.
p. Plot Mg_u and Mg_c vs. Y on linear graph paper for data and model.
q. Discuss the results.

Growth response of tobacco in North Carolina

t wk	Y Mg ha^{-1}	N_c	P_c	K_c g kg^{-1}	Ca_c	Mg_c
19.0	—	—	—	—	—	—
22.0	0.046	46	—	56	14	4.8
23.0	0.28	46	3.6	62	17	4.0
24.0	0.56	48	3.6	62	17	5.9
26.0	1.76	30	2.3	48	10	3.8
28.0	3.33	18	2.1	36	8.3	2.7
30.0	3.87	15	2.1	33	7.7	2.6
32.0	4.29	15	2.1	30	8.3	2.6

Data adapted from Raper and McCants (1967).

3.7 Shih and Gascho (1980) measured growth of sugarcane (*Saccharum* spp.) on Pahokee muck (Typic medisaprist) in Florida for two row spacings (0.5 and 1.5 m). Data are given in the table below.

a. Plot dry matter Y vs. time since Jan. 1, t, for the two-row spacing on linear paper.
b. Calculate dimensionless time x and growth quantifier Q for the parameters $t_i = 20$ wk, $\mu = 26$ wk, $\sqrt{2}\sigma = 8$ wk, $c = 0.1$ wk^{-1}, $k = 5$.
c. Calculate dry matter accumulation from $Y = 7.36\, Q$ and $Y = 4.83\, Q$ for row spacing of 0.5 m and 1.5 m, respectively.
d. Draw the simulation curves on (a).
e. Leaf area index (LAI) at the end of August ($t = 34.8$ wk) was estimated as 5.06 and 3.89 for row spacing of 0.5 m and 1.5 m, respectively. Plot the yield factor A vs. LAI on linear graph paper.
f. Discuss the results.

Growth response of sugarcane in Florida

t	Y Mg ha^{-1}	
wk	0.5 m	1.5 m
21.6	3.13	0.90
26.0	16.95	7.11
30.4	21.10	12.60
34.8	34.29	26.39

Data adapted from Shih and Gascho (1980).

3.8 Overman and Scholtz (1999) reported growth response of DeKalb T1100 hybrid corn on Kershaw fine sand (fine-sand, silicieous, thermic Quartzipsamments) at Tallahassee, FL under irrigation with reclaimed water. Planting date was March 6, 1990. Plant population was 88,000 plants ha^{-1}. The N-P-K application rate was 143-13-260 kg ha^{-1} (from fertilizer and reclaimed water). Data for accumulation of dry matter and plant nutrients are listed in the table.

a. Assume the parameters $t_i = 14$ wk, $\mu = 26$ wk, $\sqrt{2}\sigma = 8$ wk, $c = 0.2$ wk^{-1}, $k = 5$. Show that Q is given by $Q = 0.326$ $\{4.50\,(\mathrm{erf}\,x + 0.678) - 2.82\,[\exp(-x^2) - 0.613]\}$. Calculate x and Q for each sampling date.

b. Assume the yield plateau is 25.0 Mg ha^{-1}. Show that dry matter Y can be estimated from $Y = 8.26\,Q$. Calculate dry matter for each sampling date.

c. Show that dry matter and plant N accumulation N_u are related by $Y/N_u = 0.0336 + 0.00473\,Y$ and that plant N uptake N_u is given by $N_u = 211\,Y/(7.1 + Y)$. Calculate plant N uptake and plant N concentration N_c for each sampling date. Plot Y, N_u and N_c vs. t on linear graph paper.

d. Show that dry matter and plant P accumulation P_u are related by $Y/P_u = 0.263 + 0.00770\,Y$ and that plant P uptake P_u is given by $P_u = 130\,Y/(34.2 + Y)$. Calculate plant P uptake and plant P concentration P_c for each sampling date. Plot Y, P_u, and P_c vs. t on linear graph paper.

e. Show that dry matter and plant K accumulation K_u, are related by $Y/K_u = 0.0128 + 0.00751\,Y$ and that plant K uptake K_u is given by $K_u = 133\,Y/(1.7 + Y)$. Calculate plant K uptake and plant K concentration K_c for each sampling date. Plot Y, K_u, and K_c vs. t on linear graph paper.

f. Show that dry matter and plant Ca accumulation Ca_u are related by $Y/Ca_u = 0.110 + 0.0314\,Y$ and that plant Ca uptake Ca_u is given by $Ca_u = 31.9\,Y/(3.5 + Y)$. Calculate plant Ca uptake and plant Ca concentration, Ca_c for each sampling date. Plot Y, Ca_u, and Ca_c vs. t on linear graph paper.

g. Show that dry matter and plant Mg accumulation Mg_u are related by $Y/Mg_u = 0.268 + 0.0432\,Y$ and that plant Mg uptake Mg_u is given by $Mg_u = 23.1\,Y/(6.2 + Y)$. Calculate plant Mg uptake and plant Mg concentration Mg_c for each sampling date. Plot Y, Mg_u, and Mg_c vs. t on linear graph paper.

h. Discuss the results. Does the model describe coupling between dry matter and plant nutrient uptake for each nutrient adequately?

Data for corn study at Tallahassee, FL (1990)

t wk	Y Mg ha^{-1}	N_u kg ha^{-1}	P_u kg ha^{-1}	K_u kg ha^{-1}	Ca_u kg ha^{-1}	Mg_u kg ha^{-1}
9.3	—	—	—	—	—	—
15.3	0.45	12.5	1.65	9.5	3.0	1.4
16.3	1.03	26.9	4.07	22.8	7.3	3.5
17.3	1.98	40.5	7.50	50.5	11.4	5.9
18.3	4.17	61.2	12.7	97.6	17.2	9.2
19.3	5.96	112	18.6	112	23.1	11.9
20.3	7.40	137	23.3	120	21.7	12.3
21.3	9.00	125	27.4	116	21.9	13.3
22.3	10.82	137	32.2	117	25.0	14.6
23.3	13.03	127	36.5	124	26.2	15.7
24.3	16.09	139	45.5	130	26.6	18.3
25.3	17.22	153	42.4	128	25.1	16.0
26.3	16.29	160	39.4	101	21.7	14.4
27.3	17.24	138	43.4	120	20.4	14.6

3.9 Fulkerson (1983) reported growth response of orchardgrass (*Dactylis glomerata* L.) in Canada. Data are adapted from Christie and McElroy (1995) and reported in the table below for dry matter Y, plant N uptake N_u, and plant N concentration N_c. Sampling time t is referenced to Jan. 1.

a. Assume the parameters $t_i = 17\,\text{wk}$, $\mu = 26\,\text{wk}$, $\sqrt{2}\sigma = 8\,\text{wk}$, $c = 0.4\,\text{wk}^{-1}$, $k = 5$. Show that Q is given by $Q = 4.57\{-1.375\,(\text{erf}\,x - 0.498) - 2.82\,[\exp(-x^2) - 0.798]\}$. Calculate x and Q for each sampling date.

b. Assume the yield at 25 wk is 5.3 Mg ha^{-1}. Show that dry matter Y can be estimated from $Y = 0.898\ Q$. Calculate dry matter for each sampling date.

c. Show that dry matter and plant N accumulation N_u up to 24.1 wk are related by $Y/N_u = 0.0151 + 0.0115\ Y$ and that plant N uptake N_u is given by $N_u = 87.0\ Y/(1.3 + Y)$. Calculate plant N uptake and plant N concentration N_c for each sampling date.

d. Discuss the results.

Growth response of orchardgrass at Guelph, Canada

t wk	Y Mg ha^{-1}	N_u kg ha^{-1}	N_c g kg^{-1}	Y/N_u kg g^{-1}
18.1	0.535	24.0	44.8	0.022
19.1	1.10	40.8	37.1	0.027
20.3	2.02	51.7	25.6	0.039
21.1	2.85	59.3	20.8	0.048
22.1	4.01	69.4	17.3	0.058
23.1	4.63	68.1	14.7	0.068
24.1	5.00	67.0	13.4	0.075
25.1	5.47	65.6	12.0	0.083
26.3	5.57	61.3	11.0	0.091
27.1	5.76	61.1	10.6	0.094
27.4	5.71	56.0	9.8	0.102
29.1	5.19	51.4	9.9	0.101

Data adapted from Fulkerson (1983) as reported by Christie and McElroy (1995).

3.10 Soltanpour (1969) reported dry matter and nutrient accumulation by potato (*Solanum tuberosum*) on a sandy loam soil at the San Luis Valley of Colorado at high altitude (7654 msl). Data are listed below for the Oromonte cultivar.

a. Plot total dry matter (excluding roots) Y plant uptake of nitrogen, N_u, and plant nitrogen concentration N_c vs. calendar time t on linear graph paper.

b. Plot N_u and Y/N_u vs. Y on linear graph paper.

c. Use linear regression (excluding the first data point) for Y/N_u vs. Y to obtain

$$\frac{Y}{N_u} = \frac{K'_n}{N_{um}} + \frac{1}{N_{um}}Y = 0.0211 + 0.00295\,Y \qquad r = 0.977$$

Draw the line from this equation and the curve from the hyperbolic equation on (b):

$$N_u = \frac{N_{um} Y}{K'_n + Y} = \frac{340 \, Y}{7.1 + Y}$$

d. Plot P_u and Y/P_u vs. Y on linear graph paper.
e. Use linear regression (excluding the first two data points) for Y/P_u vs. Y to obtain

$$\frac{Y}{P_u} = \frac{K'_p}{P_{um}} + \frac{1}{P_{um}} Y = 0.276 + 0.0163 \, Y \qquad r = 0.966$$

Draw the line from this equation and the curve from the hyperbolic equation (d):

$$P_u = \frac{P_{um} Y}{K'_p + Y} = \frac{61.4 \, Y}{16.9 + Y}$$

f. Plot K_u and Y/K_u vs. Y on linear graph paper.
g. Use linear regression (excluding the first two data points) for Y/K_u vs. Y to obtain

$$\frac{Y}{K_u} = \frac{K'_k}{K_{um}} + \frac{1}{K_{um}} Y = 0.0203 + 0.00130 \, Y \qquad r = 0.968$$

Draw the line from this equation and the curve from the hyperbolic equation (f):

$$K_u = \frac{K_{um} Y}{K'_k + Y} = \frac{770 \, Y}{15.6 + Y}$$

h. Dry matter accumulation with time can be simulated as follows. Choose the parameters $\mu = 32$ wk, $\sqrt{2}\sigma = 8$ wk, $c = 0.1$ wk^{-1}, $k = 5$, $t_i = 25.8$ wk. Dimensionless time is then given by

$$x = \frac{t - \mu}{\sqrt{2}\sigma} + \frac{\sqrt{2}\sigma c}{2} = \frac{t - 32}{8} + 0.4 = \frac{t - 28.8}{8}$$

and the growth quantifier by

$$Q = \exp(\sqrt{2}\sigma c x_i)\Big\{(1 - kx_i)[\text{erf } x - \text{erf } x_i]$$
$$- \frac{k}{\sqrt{\pi}}[\exp(-x^2) - \exp(-x_i^2)]\Big\}$$

$$= 0.741\{2.875[\text{erf } x + 0.404] - 2.821[\exp(-x^2) - 0.869]\}$$

Calculate values of Q for $t = 25.8, 27, 28, \ldots, 44$ wk. Calculate corresponding values of Y from the calibration $Y = (14.5/3.44)Q = 4.21Q$. Plot the calculated values of Y vs. t on (a).

Dry matter and plant nutrient accumulation by potato at the San Luis Valley of Colorado

t wk	Y Mg ha^{-1}	N_u kg ha^{-1}	P_u kg ha^{-1}	K_u kg ha^{-1}	N_c g kg^{-1}	P_c g kg^{-1}	K_c g kg^{-1}
21.4	—	—	—	—	—	—	—
26.7	1.2	30	2	20	25	2.0	17
27.7	1.5	60	3	40	40	2.0	27
28.7	2.1	80	7	90	38	3.2	43
29.7	3.0	110	9	120	37	3.0	40
30.7	4.5	120	14	160	27	3.1	36
31.7	6.2	170	17	230	27	2.7	37
32.7	8.0	170	18	270	21	2.2	34
33.7	9.8	200	23	320	20	2.3	33
34.7	11.9	190	24	350	16	2.0	29
35.7	13.7	210	26	330	15	1.9	24
36.7	16.7	240	31	390	14	1.9	23
37.7	16.4	260	32	400	16	2.0	24

Data adapted from Soltanpour (1969).

3.11 Mutti (1984) measured the response of corn to applied N and without irrigation on Millhopper fine sand (loamy, hyperthermic, Grossarenic Paleudults) at Gainesville, FL. Results are given in the table below. Values are for total above-ground plant. Time is calculated as Julian week from Jan. 1. Emergence occurred at 9.7 wk.

a. Plot dry matter yield Y, plant N uptake N_u, and plant N concentration N_c vs. time t on linear graph paper.

b. For the expanded growth model assume the following parameters: $\mu = 26$ wk, $\sqrt{2}\sigma = 8.0$ wk, $c = 0.2$ wk^{-1}, $k = 5$, and time of initiation $t_i = 12.0$ wk. Show that dimensionless time and the growth quantifier are given by

$$x = \frac{t - \mu}{\sqrt{2}\sigma} + \frac{\sqrt{2}\sigma c}{2} = \frac{t - 19.6}{8.0}$$

$$Q = \exp\left(\sqrt{2}\sigma c x_i\right)\left\{(1 - kx_i)[\text{erf } x - \text{erf } x_i]\right.$$

$$\left. - \frac{k}{\sqrt{\pi}}[\exp(-x^2) - \exp(-x_i^2)]\right\}$$

$$= 0.219\{5.75[\text{erf } x + 0.821] - 2.821[\exp(-x^2) - 0.406]\}$$

c. Calculate the growth quantifier for times of $t = 13, 14, 15, \ldots, 31, 32$ wk

d. Estimate yields from the equation

$$Y = \frac{5.50}{1.72} Q = 3.20Q$$

and plot these values on (a).

e. Construct the phase plots (N_u and Y/N_u vs. Y) on linear graph paper.

f. Draw the phase relations on (d) on linear graph paper from the equations

$$\frac{Y}{N_u} = \frac{K_y}{N_{um}} + \frac{1}{N_{um}} Y = 0.0463 + 0.00767 Y \qquad r = 0.7710$$

$$N_u = \frac{N_{um} Y}{K_y + Y} = \frac{130 Y}{6.04 + Y}$$

g. Estimate plant N concentration and plot on (a) from the equation

$$N_c = \frac{N_u}{Y} = \frac{130}{6.04 + Y}$$

h. Construct the phase plots (K_u and Y/K_u vs. Y) on linear graph paper.

i. Draw the phase relations on (h) on linear graph paper from the equations

$$\frac{Y}{K_u} = \frac{K_y}{K_{um}} + \frac{1}{K_{um}} Y = 0.0144 + 0.0107 Y \qquad r = 0.9771$$

$$K_u = \frac{K_{um} Y}{K_y + Y} = \frac{93.5 Y}{1.35 + Y}$$

j. Discuss the results.

Growth response of corn with applied N of 116 kg ha^{-1} and without irrigation at Gainesville, FL

t wk	Y Mg ha^{-1}	N_u kg ha^{-1}	N_c g kg^{-1}	K_u kg ha^{-1}	K_c g kg^{-1}
9.7	—	—	—	—	—
12.9	0.0875	3.2	36.6	5.3	60.6
14.9	0.341	7.5	24.9	17.9	52.5
16.7	1.07	17.9	16.7	35.5	33.2
18.9	3.02	40.6	13.4	75.0	24.8
19.7	3.26	36.5	11.2	76.9	23.6

20.9	2.93	31.2	10.6	64.3	21.9
22.7	5.69	67.8	11.9	79.6	14.0
24.7	6.20	74.3	12.0	77.1	12.4
27.1	5.53	69.5	12.6	65.7	11.9

Data adapted from Mutti (1984, table 17).

3.12 Mutti (1984) measured the response of corn to applied N and irrigation on Millhopper fine sand (loamy, hyperthermic, Grossarenic Paleudults) at Gainesville, FL. Results are given in the table below. Values are for total above-ground plant. Time is calculated as Julian week from Jan. 1. Emergence occurred at 9.7 wk.

a. Plot dry matter yield Y, plant N uptake N_u, and plant N concentration N_c vs. time t on linear graph paper.

b. For the expanded growth model assume the following parameters: $\mu = 26$ wk, $\sqrt{2}\sigma = 8.0$ wk, $c = 0.2$ wk^{-1}, $k = 5$, and time of initiation $t_i = 12.0$ wk. Show that dimensionless time and the growth quantifier are given by

$$x = \frac{t - \mu}{\sqrt{2}\sigma} + \frac{\sqrt{2}\sigma c}{2} = \frac{t - 19.6}{8.0}$$

$$Q = \exp\left(\sqrt{2}\sigma c x_i\right)\left\{(1 - kx_i)[\text{erf } x - \text{erf } x_i]\right.$$
$$\left. - \frac{k}{\sqrt{\pi}}[\exp(-x^2) - \exp(-x_i^2)]\right\}$$

$$= 0.219\{5.75[\text{erf } x + 0.821] - 2.821[\exp(-x^2) - 0.406]\}$$

c. Calculate the growth quantifier for times of $t = 13, 14, 15, \ldots, 31, 32$ wk

d. Estimate yields from the equation

$$Y = \frac{15.0}{1.72}Q = 8.72Q$$

and plot these values on (a).

e. Construct the phase plots (N_u and Y/N_u vs. Y) on linear graph paper.

f. Draw the phase relations on (d) on linear graph paper from the equations

$$\frac{Y}{N_u} = \frac{K_y}{N_{um}} + \frac{1}{N_{um}}Y = 0.0442 + 0.00586\,Y \qquad r = 0.8916$$

$$N_u = \frac{N_{um}\,Y}{K_y + Y} = \frac{170\,Y}{7.54 + Y}$$

g. Estimate plant N concentration and plot on (a) from the equation

$$N_c = \frac{N_u}{Y} = \frac{170}{7.54 + Y}$$

h. Construct the phase plots (K_u and Y/K_u vs. Y) on linear graph paper.

i. Draw the phase relations on (h) on linear graph paper from the equations

$$\frac{Y}{K_u} = \frac{K_y}{K_{um}} + \frac{1}{K_{um}} Y = 0.0167 + 0.00565 Y \qquad r = 0.9954$$

$$K_u = \frac{K_{um} Y}{K_y + Y} = \frac{177 \, Y}{2.95 + Y}$$

j. Discuss the results.

Growth response of corn with applied N of 116 kg ha^{-1} and under irrigation at Gainesville, FL

t wk	Y Mg ha^{-1}	N_u kg ha^{-1}	N_c g kg^{-1}	K_u kg ha^{-1}	K_c g kg^{-1}
9.7	—	—	—	—	—
12.9	0.106	3.7	34.9	6.4	60.4
14.9	0.348	7.9	22.7	17.4	50.0
16.7	1.54	38.4	24.9	58.2	37.8
18.9	5.64	59.6	10.6	127	22.5
19.7	7.40	59.7	8.1	128	17.3
20.9	8.29	82.8	10.0	121	14.6
22.7	12.88	119	9.2	152	11.8
24.7	15.36	113	7.4	149	9.7
27.1	14.58	131	9.0	141	9.7

Data adapted from Mutti (1984, table 20).

3.13 Mutti (1984) measured the response of corn to applied N and without irrigation on Millhopper fine sand (loamy, hyperthermic, Grossarenic Paleudults) at Gainesville, FL. Results are given in the table below. Values are for total above-ground plant. Time is calculated as Julian week from Jan. 1. Emergence occurred at 9.7 wk.

a. Plot dry matter yield Y, plant N uptake N_u, and plant N concentration N_c vs. time t on linear graph paper.

b. For the expanded growth model assume the following parameters: $\mu = 26$ wk, $\sqrt{2}\sigma = 8.0$ wk, $c = 0.2$ wk^{-1}, $k = 5$, and time of initiation $t_i = 12.0$ wk. Show that dimensionless time and the growth quantifier are given by

$$x = \frac{t - \mu}{\sqrt{2\sigma}} + \frac{\sqrt{2}\sigma c}{2} = \frac{t - 19.6}{8.0}$$

$$Q = \exp\left(\sqrt{2}\sigma cx_i\right)\left\{(1 - kx_i)[\text{erf } x - \text{erf } x_i] - \frac{k}{\sqrt{\pi}}[\exp(-x^2)\right.$$

$$\left. - \exp(-x_i^2)]\right\}$$

$$= 0.219\{5.75[\text{erf } x + 0.821] - 2.821[\exp(-x^2) - 0.406]\}$$

c. Calculate the growth quantifier for times of $t = 13,\ 14,\ 15,\ \ldots,\ 31,\ 32$ wk.
d. Estimate yields from the equation

$$Y = \frac{8.60}{1.72}Q = 5.00Q$$

and plot these values on (a).
e. Construct the phase plots (N_u and Y/N_u vs. Y) on linear graph paper.
f. Draw the phase relations on (d) on linear graph paper from the equations

$$\frac{Y}{N_u} = \frac{K_y}{N_{um}} + \frac{1}{N_{um}}Y = 0.0345 + 0.00654\,Y \qquad r = 0.8642$$

$$N_u = \frac{N_{um}\,Y}{K_y + Y} = \frac{153\,Y}{5.28 + Y}$$

g. Estimate plant N concentration and plot on (a) from the equation

$$N_c = \frac{N_u}{Y} = \frac{153}{5.28 + Y}$$

h. Construct the phase plots (K_u and Y/K_u vs. Y) on linear graph paper.
i. Draw the phase relations on (h) on linear graph paper from the equations

$$\frac{Y}{K_u} = \frac{K_y}{K_{um}} + \frac{1}{K_{um}}Y = 0.0179 + 0.00845\,Y \qquad r = 0.9741$$

$$K_u = \frac{K_{um}\,Y}{K_y + Y} = \frac{118\,Y}{2.12 + Y}$$

j. Discuss the results.

Growth response of corn with applied N of 401 kg ha^{-1} and without irrigation at Gainesville, FL

t wk	Y Mg ha^{-1}	N_u kg ha^{-1}	N_c g kg^{-1}	K_u kg ha^{-1}	K_c g kg^{-1}
9.7	—	—	—	—	—
12.9	0.0958	4.5	47.0	5.4	55.4
14.9	0.626	18.7	29.9	27.2	47.8
16.7	1.77	31.2	17.6	42.9	24.2
18.9	4.46	65.5	14.7	101	22.6
19.7	4.58	61.5	13.4	83.3	18.2
20.9	4.66	58.4	12.5	75.7	16.2
22.7	6.63	93.0	14.0	88.6	13.4
24.7	7.55	92.0	12.2	89.1	11.8
27.1	7.20	105	14.6	92.2	12.8

Data adapted from Mutti (1984, table 18).

3.14 Mutti (1984) measured the response of corn to applied N and irrigation on Millhopper fine sand (loamy, hyperthermic, Grossarenic Paleudults) at Gainesville, FL. Results are given in the table below. Values are for total above-ground plant. Time is calculated as Julian week from Jan. 1. Emergence occurred at 9.7 wk.

a. Plot dry matter yield Y, plant N uptake N_u, and plant N concentration N_c vs. time t on linear graph paper.

b. For the expanded growth model assume the following parameters: $\mu = 26$ wk, $\sqrt{2}\sigma = 8.0$ wk, $c = 0.2$ wk^{-1}, $k = 5$, and time of initiation $t_i = 12.0$ wk. Show that dimensionless time and the growth quantifier are given by

$$x = \frac{t - \mu}{\sqrt{2}\sigma} + \frac{\sqrt{2}\sigma c}{2} = \frac{t - 19.6}{8.0}$$

$$Q = \exp\left(\sqrt{2}\sigma c x_i\right)\left\{(1 - kx_i)[\mathrm{erf}\, x - \mathrm{erf}\, x_i]\right.$$
$$\left. - \frac{k}{\sqrt{\pi}}[\exp(-x^2) - \exp(-x_i^2)]\right\}$$
$$= 0.219\{5.75[\mathrm{erf}\, x + 0.821] - 2.821[\exp(-x^2) - 0.406]\}$$

c. Calculate the growth quantifier for times of $t = 13, 14, 15, \ldots, 31, 32$ wk.

d. Estimate yields from the equation

$$Y = \frac{21.0}{1.72}Q = 12.2Q$$

and plot these values on (a).

e. Construct the phase plots (N_u and Y/N_u vs. Y) on linear graph paper.

f. Draw the phase relations on (d) on linear graph paper from the equations

$$\frac{Y}{N_u} = \frac{K_y}{N_{um}} + \frac{1}{N_{um}} Y = 0.0310 + 0.00255\,Y \qquad r = 0.9743$$

$$N_u = \frac{N_{um}\,Y}{K_y + Y} = \frac{392\,Y}{12.2 + Y}$$

g. Estimate plant N concentration and plot on (a) from the equation

$$N_c = \frac{N_u}{Y} = \frac{392}{12.2 + Y}$$

h. Construct the phase plots (K_u and Y/K_u vs. Y) on linear graph paper.

i. Draw the phase relations on (h) on linear graph paper from the equations

$$\frac{Y}{K_u} = \frac{K_y}{K_{um}} + \frac{1}{K_{um}} Y = 0.0197 + 0.00365\,Y \qquad r = 0.9965$$

$$K_u = \frac{K_{um}\,Y}{K_y + Y} = \frac{274\,Y}{5.39 + Y}$$

j. Discuss the results.

Growth response of corn with applied N of 401 kg ha^{-1} and under irrigation at Gainesville, FL

t wk	Y Mg ha^{-1}	N_u kg ha^{-1}	N_c g kg^{-1}	K_u kg ha^{-1}	K_c g kg^{-1}
9.7	—	—	—	—	—
12.9	0.0936	4.2	44.9	5.2	55.6
14.9	0.651	19.0	29.2	28.3	43.5
16.7	2.46	63.4	25.8	80.5	32.7
18.9	8.30	152	18.3	170	20.5
19.7	10.75	169	16.7	190	17.8
20.9	13.69	188	13.7	196	14.3
22.7	16.08	211	13.1	192	11.9
24.7	22.39	267	11.9	226	10.1
27.1	22.00	268	12.2	221	10.0

Data adapted from Mutti (1984, table 20).

3.15 We now examine results from the corn study by Mutti (1984) at Gainesville, FL discussed in Exercises 3.11 through 3.14 above. In particular, we are interested in the dependence of yield factor A on applied nitrogen and water availability W, which consists of irrigation and/or rainfall.

a. Plot A vs. W from the table below on linear graph paper.
b. Assume a linear relationship between A and W and obtain the equations

$$N = 116 \text{ kg ha}^{-1} \qquad A = 0.272W - 1.64$$

$$= 0.272(W - 6.0) \approx 0.26(W - 5.0)$$

$$N = 401 \text{ kg ha}^{-1} \qquad A = 0.355W - 1.31$$

$$= 0.355(W - 3.7) \approx 0.38(W - 5.0)$$

c. Compare the values of parameter A for rainfed vs. irrigated treatments. Are the ratios similar for the two nitrogen levels?
d. What is the significance of the intercept 5.00 cm in the response equations?
e. Does the assumption of uniform values of the parameters $\mu, \sigma, c, k,$ and t_i in the expanded growth model for the different irrigation and Note that this assigns dependence of yields on these treatments to the yield factor A.

Dependence of model parameter A on water availability W for corn study at Gainesville, FL

Applied N kg ha^{-1}	Irrigation	A Mg ha^{-1}	W cm
116	No	3.20	17.8
	Yes	8.72	38.1
401	No	5.00	17.8
	Yes	12.2	38.1

Water availability is estimated from Mutti (1984, fig. 1) for the period 40 through 100 days after plant emergence.

3.16 Tolk et al. (1998) measured response of corn to irrigation on three soils in field lysimeters at Bushland, TX. The soils were Amarillo sandy loam (fine-loamy, mixed, superactive, thermic Aridic Paleustalf), Pullman clay loam (fine, mixed, thermic Torrertic Paleustoll), and Ulysses clay loam (fine-silty, mixed, superactive, mesic Aridic Haplustoll). This analysis focuses on data for 1996 (listed in the table below), where irrigation ranged from 12.8 to 63.9 cm. Dry matter yields are given for grain and total above-ground plant. Applied N was 180 kg ha^{-1}.

a. Plot dry matter Y vs. evapotranspiration, ET, for all three soils on linear graph paper.

b. Since the data show curvilinear behavior, calculate curves to plot on (a) from the equation

$$Y = A\{1 - \exp[-c(ET - b)]\} = A\{1 - \exp[-0.0345(ET - 12.7)]\}$$

where A(grain) $= 10.2$ and A(total) $= 20.0$ Mg ha^{-1}. Since we expect ET ≤ 100 cm, use this as the upper limit for the horizontal axis.

c. Show that for ET $= 35$ cm the estimated yield is about 50% of projected maximum.

d. Show that for ET $= 60$ cm the estimated yield is about 80% of projected maximum.

e. Discuss the results. Does it appear reasonable to group data for the three soils together for model purposes? What is the significance of this assumption for the plant system?

Response of accumulation of dry matter to evapotranspiration for corn on three soils in 1996 at Bushland, TX

Soil	Evapotranspiration, cm	Dry matter, Mg ha^{-1}	
		Grain	Total
Amarillo	34.2	5.36	9.83
sandy loam	43.6	6.70	12.48
	49.6	7.16	13.93
	53.4	7.36	14.26
Pullman	32.8	4.99	9.83
clay loam	44.6	6.84	13.08
	52.7	7.90	15.20
	61.0	8.04	15.90
Ulysses	40.1	6.55	12.93
clay loam	50.7	7.85	15.19
	56.9	8.08	15.70
	61.7	8.47	16.48

Data adapted from Tolk et al. (1998).

3.17 Hammond and Kirkham (1949) reported growth data for soybeans (total plant) on Webster and Clarion soils at Ames, IA. Results are given in the table below.

a. Plot dry matter Y vs. calendar time t for both soils on linear graph paper.

b. Assume the parameters $t_i = 25$ wk, $\mu = 32.0$ wk, $\sqrt{2}\sigma = 8.0$ wk, $c = 0.1$ wk, $k = 5$. Calculate dimensionless time and the growth quantifier from

$$x = \frac{t - 28.8}{8}$$

$$Q = 0.684\{3.375[\text{erf } x + 0.498] - 2.821[\exp(-x^2) - 0.798]\}$$

c. Estimate plant dry matter from the equations and draw the model curves on (a).

Webster soil: $\quad Y = \left(\dfrac{18.5}{3.31}\right)Q = 5.59Q$

Clarion soil: $\quad Y = \left(\dfrac{12.5}{3.31}\right)Q = 3.78Q$

d. Discuss fit of the model to the data.

Growth data for soybeans on Webster and
Clarion soils at Ames, IA

Calendar time wk	Dry matter, g plant^{-1}	
	Webster soil	Clarion soil
21.4	—	—
24.6	0.34	0.30
25.9	0.68	0.62
27.3	1.84	1.45
27.9	2.40	1.99
28.9	4.18	2.83
29.9	5.72	3.78
30.9	7.80	5.42
31.9	10.22	7.62
32.9	13.93	9.19
33.9	16.09	10.47
34.9	18.56	11.42
35.9	20.88	15.48

Data adapted from Hammond and Kirkham (1949).

3.18 Stanley (1994) studied the response of "Tifton-9" bahiagrass to harvest interval and applied N on Dothan loamy fine sand (fine-loamy, siliceous, thermic Plinthic Kandiudult) at Quincy, FL. The experiment has been analyzed by Overman and Stanley (1998). Some of the results are given in the tables below.

a. Plot dry matter yield Y vs. applied nitrogen N on linear graph paper.

b. Draw model estimates on (a) from the logistic equation

$$Y = \frac{21.5}{1 + \exp(0.72 - 0.0050N)}$$

c. Plot dry matter yield Y vs. harvest interval Δt on linear graph paper.

d. Calculate yield standardized for applied N, Y^*, for each harvest interval from the equation

$$Y^* = Y \exp(0.72 - 0.0050N)$$

e. Calculate yield standardized for harvest interval Y^{**} for each harvest interval from the equation

$$Y^{**} = Y^* \exp(\gamma \Delta t) = Y * \exp(0.100 \Delta t)$$

f. Perform linear regression on Y^{**} vs. Δt to obtain the linear equation

$$Y^{**} = \alpha + \beta \Delta t = 6.75 + 5.85 \Delta t \qquad r = 0.9963$$

g. Plot Y^{**} vs. Δt for data and model on linear graph paper.

h. Plot Y vs. Δt on (c) for applied N of 336 kg ha^{-1} from the equation

$$Y = \frac{(6.75 + 5.85 \Delta t) \exp(-0.100 \Delta t)}{1 + \exp(0.72 - 0.0050N)}$$

i. Show that harvest interval for peak yield Δt_p is given by

$$\Delta t_p = \frac{1}{\gamma} - \frac{\alpha}{\beta} = 10.0 - 1.2 = 8.8 \text{ wk}$$

j. Discuss agreement between the model and data for this system.

Yield response of "Tifton-9"
bahiagrass to applied N for a harvest
interval of 4 wk at Quincy, FL

Applied N kg ha^{-1}	Dry matter Mg ha^{-1}
0	7.45
84	8.73

168	11.00
336	15.63
672	20.00

Data adapted from Stanley (1994).

Yield response of "Tifton-9"
bahiagrass to harvest interval
for applied N of 336 kg ha^{-1} at
Quincy, FL

Harvest interval wk	Dry matter Mg ha^{-1}
1	6.92
2	9.82
4	15.35
8	18.95
16	14.30

Data adapted from Stanley (1994).

3.19 Beaty et al. (1963) conducted a study of response to applied N and harvest interval of pensacola bahiagrass on Red Bay sandy loam (fine-loamy, kaolinitic, thermic Rhodic Kandiudults) at Americus, GA. Data are given in Table 1 below.

   a. Plot dry matter yield Y vs. applied nitrogen N for each harvest interval Δt on linear graph paper.

   b. Construct model curves on (a) from the logistic equation

   $$Y = \frac{A}{1 + \exp(1.35 - 0.0135N)}$$

   using values for parameter A from Table 2.

   c. Calculate standardized values A^* for each harvest interval from the equation

   $$A^* = A\exp(\gamma\Delta t) = A\exp(0.100\Delta t)$$

   d. Perform linear regression on A^* vs. Δt to obtain the equation

   $$A^* = \alpha + \beta\Delta t = 3.00 + 1.99\Delta t \qquad r = 0.9959$$

   e. Draw A and A^* vs. Δt on linear graph paper.

f. Draw the curve for A vs. Δt on (e) from the linear-exponential equation

$$A = (3.00 + 1.99\Delta t)\exp(-0.100\Delta t)$$

Show that the peak harvest interval Δt_p is given by

$$\Delta t_p = \frac{1}{\gamma} - \frac{\alpha}{\beta} = \frac{1}{0.1} - \frac{3.00}{1.99} = 8.5 \text{ wk}$$

g. Estimate yield values in Table 3 and draw the curves on (a) from the equation

$$\widehat{Y} = \frac{(3.00 + 1.99\Delta t)\exp(-0.100\Delta t)}{1 + \exp(1.35 - 0.0135N)}$$

h. Construct a scatter diagram of estimated vs. measured yields on linear graph paper. Show that the correlation is given by

$$\widehat{Y} = 0.077 + 0.969Y \qquad r = 0.9889$$

Table 1   Response of Pensacola bahiagrass to applied N and harvest interval at Americus, GA

Applied N kg ha^{-1}	Dry matter Mg ha^{-1} Harvest interval, wk				
	1	2	3	4	6
0	1.08	1.35	1.50	1.47	1.58
56	1.77	2.16	2.33	2.42	2.75
112	2.26	2.86	3.44	3.46	4.01
224	3.92	4.78	6.19	6.28	7.18

Data adapted from Beaty et al. (1963).

Table 2   Dependence of parameter A and A^* on harvest interval

Δt wk	A Mg ha^{-1}	A^* Mg ha^{-1}	$\widehat{A}^*$ Mg ha^{-1}
1	4.50	4.97	4.50
2	5.50	6.72	5.70
3	7.10	9.58	6.65
4	7.20	10.74	7.35
6	8.20	14.94	8.20

Table 3 Estimated response of Pensacola bahiagrass to
applied N and harvest interval at Americus, GA

N	$\hat{Y}$				
	Mg ha^{-1}				
	Δt, wk				
kg ha^{-1}	1	2	3	4	6
0	0.93	1.17	1.37	1.51	1.69
56	1.60	2.03	2.37	2.61	2.92
112	2.43	3.08	3.59	3.97	4.43
224	3.79	4.80	5.60	6.19	6.90

3.20 Crowder et al. (1955) reported on the effect of harvest interval of a
grass mixture (oats, ryegrass, crimson clover) on Davidson clay loam
(Fine, kaolinitic, thermic Rhodic Kandiudults) at Experiment, GA.
Data are given in the table below, which represent seasonal averages
over three seasons and three rates of top dressing with applied N.
  a. Plot Y, N_u, and N_c vs. Δt on linear graph paper.
  b. Calculate standardized yield Y^* and plant N uptake N^* for each
     harvest interval from the equations

$$Y^* = Y \exp(\gamma \Delta t)$$

$$N_u = N_u \exp(\gamma \Delta t)$$

  using the value $\gamma = 0.08$.
  c. Plot Y^* and N_u^* vs. Δt on linear graph paper.
  d. Plot the model lines on (c) from the equations

$$Y^* = 1.32 + 1.48\Delta t \qquad r = 0.9994$$

$$N_u^* = 132 + 25.9\Delta t \qquad r = 0.9935$$

  e. Plot the model curves on (a) from the equations

$$Y = (1.32 + 1.48\Delta t)\exp(-0.08\Delta t)$$

$$N_u = (132 + 25.9\Delta t)\exp(-0.08\Delta t)$$

$$N_c = \frac{132 + 25.9\Delta t}{1.32 + 1.48\Delta t}$$

  f. Estimate the harvest interval for maximum yield and maximum
     plant N uptake.
  g. Discuss the results.

Dependence of dry matter yield Y, plant N uptake N_u and plant N concentration N_c on harvest interval Δt for the grass mixture of oats, ryegrass, and crimson clover at Experiment, GA

Harvest interval wk	Dry matter Mg ha^{-1}	Plant N uptake kg ha^{-1}	Plant N concentration g kg^{-1}
2	3.74	151.0	40.4
4	5.12	178.5	34.9
8	6.95	177.2	25.5

Data adapted from Crowder et al. (1955).

3.21 Liu et al. (1997) studied treatment of swine lagoon effluent by the overland flow method on Marvyn loamy sand (fine-loamy, kaolinitic, thermic Typic Kanhapludult) with a 10% slope at Auburn, AL. Vegetation was hybrid bermudagrass (*Cynodon dactylon* (L.) Pers. cv. Russell) overseeded with annual ryegrass (*Lolium multiflorum* Lam.). This analysis will focus on control, NH$_4$NO$_3$ at 560 kg ha^{-1} of N, and swine effluent at 560 kg ha^{-1} of N. Effluent added 71 kg ha^{-1} of P. Data are given in Tables 1, 2, and 3 for 1993 for harvest times from Jan. 1, t, dry matter increment ΔY, cumulative dry matter Y, plant N increment ΔN, cumulative plant N, N, plant P increment ΔP, and cumulative plant P, P, and cumulative fractions of dry matter F_y, plant N, F_n, and plant P, F_p.

a. Plot Y, N_u, and P_u vs. t for the three treatments on linear graph paper.

b. Plot F_y vs. t for the control treatment on probability paper. Show that the mean and standard deviation for the distribution are given by $\mu = 26.0$ wk and $\sigma = 11$ wk, respectively. Note that σ reflects the combined production of the two grasses.

c. Use math tables for the error function (erf) to calculate F vs. t from the equation

$$F = \frac{1}{2}\left[1 + \text{erf}\left(\frac{t - 26.0}{15.55}\right)\right]$$

d. Use results of (c) to calculate Y vs. t from the equation

$$Y = A_y F$$

where A_y(control) = 9.20 Mg ha^{-1}, A_y(fertilizer) = 15.00 Mg ha^{-1}, and A_y(effluent) = 15.00 Mg ha^{-1}. Plot the curves on (a).

e. Use results of (c) to calculate N_u vs. t from the equation

$$N_u = A_n F$$

where A_n(control) $= 180$ kg ha^{-1}, A_n(fertilizer) $= 350$ kg ha^{-1}, and A_n(effluent) $= 330$ kg ha^{-1}. Plot the curves on (a).

f. Use results of (c) to calculate P_u vs. t from the equation

$$P_u = A_p F$$

where A_p(control) $= 17.5$ kg ha^{-1}, A_p(fertilizer) $= 32.0$ kg ha^{-1}, and A_p(effluent) $= 45.0$ kg ha^{-1}. Plot the curves on (a).

g. Discuss the results. Note that the harvest interval Δt was not constant.

Table 1 Plant accumulation of dry matter, plant N, and plant P with time for the control treatment of the overland flow study at Auburn, AL (1993)

t wk	ΔY Mg ha^{-1}	Y Mg ha^{-1}	F_y	ΔN_u kg ha^{-1}	N_u kg ha^{-1}	F_n	ΔP_u kg ha^{-1}	P_u kg ha^{-1}	F_p
—		0	0		0	0		0	0
	0.70			10.8			1.1		
10.0		0.70	0.076		10.8	0.060		1.1	0.063
	0.82			15.3			2.0		
15.9		1.52	0.165		26.1	0.145		3.1	0.177
	1.68			24.2			2.4		
22.0		3.20	0.348		50.3	0.279		5.5	0.314
	1.84			36.1			3.5		
30.1		5.04	0.548		86.4	0.480		9.0	0.514
	1.64			42.5			4.3		
34.1		6.68	0.726		128.9	0.716		13.3	0.760
	1.46			30.5			3.2		
39.0		8.14	0.885		159.4	0.886		16.5	0.943
—		9.20	1.000		180.0	1.000		17.5	1.000

Harvest data adapted from Liu et al. (1997, table 3).

Table 2 Plant accumulation of dry matter, plant N, and plant P with time for the fertilizer treatment of the overland flow study at Auburn, AL (1993)

t wk	ΔY Mg ha^{-1}	Y Mg ha^{-1}	F_y	ΔN_u kg ha^{-1}	N_u kg ha^{-1}	F_n	ΔP_u kg ha^{-1}	P_u kg ha^{-1}	F_p
—		0	0		0	0		0	0
	1.53			29.1			2.3		
10.0		1.53	0.102		29.1	0.083		2.3	0.072
	1.15			27.7			2.8		
15.9		2.68	0.179		56.8	0.162		5.1	0.159

t wk	ΔY Mg ha^{-1}	Y Mg ha^{-1}	F_y	ΔN_u kg ha^{-1}	N_u kg ha^{-1}	F_n	ΔP_u kg ha^{-1}	P_u kg ha^{-1}	F_p
	2.39			47.1			4.1		
22.0		5.07	0.338		103.9	0.297		9.2	0.288
	3.19			71.8			6.7		
30.1		8.26	0.551		175.7	0.502		15.9	0.497
	3.41			101.6			10.2		
34.1		11.67	0.778		277.3	0.792		26.1	0.816
	1.81			46.3			4.3		
39.0		13.48	0.899		323.6	0.925		30.4	0.950
—		15.00	1.000		350.0	1.000		32.0	1.000

Harvest data adapted from Liu et al. (1997, table 3).

Table 3 Plant accumulation of dry matter, plant N, and plant P with time for the effluent treatment of the overland flow study at Auburn, AL (1993)

t wk	ΔY Mg ha^{-1}	Y Mg ha^{-1}	F_y	ΔN_u kg ha^{-1}	N_u kg ha^{-1}	F_n	ΔP_u kg ha^{-1}	P_u kg ha^{-1}	F_p
—		0	0		0	0		0	0
	2.05			37.9			5.1		
10.0		2.05	0.137		37.9	0.115		5.1	0.113
	2.30			52.2			8.7		
15.9		4.35	0.290		90.1	0.273		13.8	0.307
	1.89			30.0			3.8		
22.0		6.24	0.416		120.1	0.364		17.6	0.391
	2.46			55.8			6.6		
30.1		8.70	0.580		175.9	0.533		24.2	0.538
	2.90			86.7			9.9		
34.1		11.60	0.773		262.6	0.796		34.1	0.758
	1.76			40.8			4.9		
39.0		13.36	0.891		303.4	0.919		39.0	0.867
—		15.00	1.000		330.0	1.000		45.0	1.000

Harvest data adapted from Liu et al. (1997, table 3).

3.22 Wagger and Cassel (1993) reported response of corn to irrigation and tillage method on Hiwassee clay loam (clayey, kaolinitic, thermic Rhodic Kanhapludult) at Salisbury, NC. Data for 1986 are listed in the table below. Rainfall is estimated for the period April through July. Water availability includes irrigation + rainfall. Applied N was 224 kg ha^{-1}.

  a. Plot dry matter yield Y for grain and silage vs. water availability W for conventional tillage treatment, CT, and no-tillage, NT, on linear graph paper.

b. Perform linear regression on Y vs. W for grain and silage and for both tillage treatments to obtain slope and intercept parameters and correlation coefficients.

c. Draw the estimation lines on (a) from the equations

Grain: $Y = 0.21(W - 5.0)$

Silage: $Y = 0.42(W - 5.0)$

d. Explain the significance of the intercept of 5 cm on the W axis.

e. Does the linear relationship between Y and W seem reasonable or would you expect an upper limit to dry matter as water availability increases to higher levels?

Response of corn grain and silage to irrigation and tillage (conventional and no-tillage) at Salisbury, NC (1986)

Tillage	Irrigation cm	Water cm	Grain yield Mg ha^{-1}	Silage yield Mg ha^{-1}
Conventional	0	13.7	1.65	4.91
	19.1	32.8	6.07	11.37
	40.0	53.7	10.62	20.28
No-tillage	0	13.7	0.98	3.52
	17.8	31.5	5.07	9.88
	41.5	55.2	10.50	21.87

Data adapted from Wagger and Cassel (1993, fig. 1 and tables 1, 3, and 4).

3.23 Staley and Perry (1995) reported response of corn to applied N for silage production on Gilpin silt loam (fine-loamy, mixed, mesic Typic Hapludult) at Cool Ridge, WV. Average results for the period 1984–1986 for conventional tillage and no-till are given in the table below.

a. Plot plant dry matter Y, plant N uptake N_u, and plant N concentration N_c vs. applied nitrogen N on linear graph paper for both tillage methods.

b. Draw the model curves on (a) from the logistic equations

$$Y = \frac{A}{1 + \exp(b - cN)} = \frac{12.5}{1 + \exp(-0.60 - 0.0200N)}$$

$$N_u = \frac{A_n}{1 + \exp(b' - cN)} = \frac{170}{1 + \exp(0.65 - 0.0200N)}$$

$$N_c = \frac{A_n}{A}\left[\frac{1 + \exp(b - cN)}{1 + \exp(b' - cN)}\right] = 13.6\left[\frac{1 + \exp(-0.60 - 0.0200N)}{1 + \exp(0.65 - 0.0200N)}\right]$$

c. Construct the phase plot (Y and N_c vs. N_u) on linear graph paper for both tillage methods.

d. Draw the phase relations on (c) from the equations

$$Y = \frac{Y_m N_u}{K' + N_u} = \frac{17.5 N_u}{68.3 + N_u}$$

$$N_c = \frac{K_n}{Y_m} + \frac{1}{Y_m} N_u = 3.90 + 0.0571 N_u$$

e. Discuss the fit of the model to the data for dry matter and plant N. Do the data for the two tillage methods appear to follow the same equations?

Response of corn silage to applied N and tillage at Cool Ridge, WV (1984–1986)

Tillage	Applied N kg ha^{-1}	Plant yield Mg ha^{-1}	Plant N uptake kg ha^{-1}	Plant N concentration g kg^{-1}
Conventional	0	7.9	58.7	7.4
	56	11.4	104	9.1
	112	11.7	134	11.5
	224	12.2	165	13.5
No tillage	0	7.8	65.1	8.4
	56	11.4	106	9.3
	112	12.1	140	11.6
	224	12.8	175	13.7

Data adapted from Staley and Perry (1995, table 4).

3.24 Cassel (1988) reported on response of corn to irrigation with conventional tillage on Wagram loamy sand (loamy, siliceous, thermic Typic Paleudult) at Clayton, NC. Data are given in the table below for 1979. Time is calculated from Jan. 1, with planting on April 18, 1979.

a. Plot dry matter Y vs. calendar time t for both treatments on linear graph paper.

b. Assume the following parameters for the growth model: $\mu = 26.0$ wk, $\sqrt{2}\sigma = 8.0$ wk, $c = 0.2$ wk^{-1}, $k = 5$. Assume time of initiation of significant growth is $t_i = 21$ wk. Calculate dimensionless time x and the growth quantifier Q from

$$x = \frac{t - \mu}{\sqrt{2}\sigma} + \frac{\sqrt{2}\sigma c}{2}$$

$$Q = \exp\left(\sqrt{2}\sigma c x_i\right)\left\{(1 - kx_i)[\text{erf } x - \text{erf } x_i]\right.$$
$$\left. - \frac{k}{\sqrt{\pi}}[\exp(-x^2) - \exp(-x_i^2)]\right\}$$

c. Calculate dry matter vs. time from the model equations (calibrated at $t = 28$ wk)

Nonirrigated: $Y = AQ = \left(\dfrac{6.00}{2.49}\right)Q = 2.41Q$

Irrigated: $Y = AQ = \left(\dfrac{6.50}{2.49}\right)Q = 2.61Q$

d. Plot the calculated values from (c) on (a).
e. Discuss the results. Does the model give a reasonable description of the data? Does the effect of irrigation appear to be adequately accounted for in the A parameter?

Growth response of corn to irrigation with conventional tillage at Clayton, NC (1979)

Calendar time	Dry matter Mg ha^{-1}	
wk	Nonirrigated	Irrigated
15.6	—	—
20.9	0.16	0.19
21.9	0.33	0.32
23.3	1.68	1.48
24.7	2.53	2.54
26.9	5.89	5.30
28.6	6.52	7.67

Data adapted from Cassel (1988, table 9).

3.25 Cassel (1988) reported on response of corn to irrigation with conventional tillage on Wagram loamy sand (loamy, siliceous, thermic Typic Paleudult) at Clayton, NC. Data are given in the table below for 1980. Time is calculated from Jan. 1, with planting on April 19, 1980.
a. Plot dry matter Y vs. calendar time t for both treatments on linear graph paper.
b. Assume the following parameters for the growth model: $\mu = 26.0$ wk, $\sqrt{2}\sigma = 8.0$ wk, $c = 0.2$ wk^{-1}, $k = 5$. Assume time of initiation of significant growth is $t_i = 20$ wk. Calculate dimensionless time x and the growth quantifier Q:

$$x = \frac{t - \mu}{\sqrt{2\sigma}} + \frac{\sqrt{2\sigma c}}{2}$$

$$Q = \exp\left(\sqrt{2\sigma c x_i}\right)\left\{(1 - kx_i)[\text{erf } x - \text{erf } x_i]\right.$$
$$\left. - \frac{k}{\sqrt{\pi}}[\exp(-x^2) - \exp(-x_i^2)]\right\}$$

c. Calculate dry matter vs. time from the model equations (calibrated at $t = 27$ wk)

Nonirrigated: $Y = AQ = \left(\dfrac{5.00}{2.36}\right)Q = 2.12Q$

Irrigated: $Y = AQ = \left(\dfrac{9.00}{2.36}\right)Q = 3.81Q$

d. Plot the calculated values from (c) on (a).
e. Discuss the results. Does the model give a reasonable description of the data? Does the effect of irrigation appear to be adequately accounted for in the A parameter?

Growth response of corn to irrigation with conventional tillage at Clayton, NC (1980)

Calendar time wk	Dry matter wk Mg ha^{-1}	
	NonIrrigated	Irrigated
15.6	—	—
18.3	0.01	0.01
20.0	0.15	0.17
21.4	0.65	0.75
23.4	2.12	3.14
25.6	4.15	6.74
27.0	4.99	7.98
29.0	5.92	12.52

Data adapted from Cassel (1988, table 10).

3.26 Cassel (1988) reported on response of corn to irrigation with conventional tillage on Wagram loamy sand (loamy, siliceous, thermic Typic Paleudult) at Clayton, NC. Data are given in the table below for 1981. Time is calculated from Jan. 1, with planting on April 16, 1981.

a. Plot dry matter Y vs. calendar time t for both treatments on linear graph paper.

b. Assume the following parameters for the growth model: $\mu = 26.0$ wk, $\sqrt{2}\sigma = 8.0$ wk, $c = 0.2$ wk^{-1}, $k = 5$. Assume time of initiation of significant growth is $t_i = 21$ wk. Calculate dimensionless time x and the growth quantifier Q:

$$x = \frac{t - \mu}{\sqrt{2}\sigma} + \frac{\sqrt{2}\sigma c}{2}$$

$$Q = \exp\left(\sqrt{2}\sigma c x_i\right)\left\{(1 - kx_i)[\text{erf } x - \text{erf } x_i]\right.$$
$$\left. - \frac{k}{\sqrt{\pi}}[\exp(-x^2) - \exp(-x_i^2)]\right\}$$

c. Calculate dry matter vs. time from the model equations (calibrated at $t = 30$ wk)

Nonirrigated: $\quad Y = AQ = \left(\dfrac{9.00}{3.05}\right)Q = 2.95Q$

Irrigated: $\quad\quad Y = AQ = \left(\dfrac{17.00}{3.05}\right)Q = 5.57Q$

d. Plot the calculated values from (c) on (a).
e. Discuss the results. Does the model give a reasonable description of the data? Does the effect of irrigation appear to be adequately accounted for in the A parameter?

Growth response of corn to irrigation with conventional tillage at Clayton, NC (1981)

Calendar time wk	Dry matter Mg ha^{-1}	
	Nonirrigated	Irrigated
15.3	—	—
20.9	0.26	0.39
21.9	1.12	1.22
23.9	4.44	5.41
26.0	5.11	8.50
28.0	7.62	12.75
30.0	9.13	18.43
31.9	8.84	20.81

Data adapted from Cassel (1988, table 11).

3.27 Cassel (1988) reported on response of corn to irrigation with subsoiling on Wagram loamy sand (loamy, siliceous, thermic Typic

Paleudult) at Clayton, NC. Data are given in the table below for 1979. Time is calculated from Jan. 1, with planting on April 18, 1979.

a. Plot dry matter Y vs. calendar time t for both treatments on linear graph paper.

b. Assume the following parameters for the growth model: $\mu = 26.0$ wk, $\sqrt{2}\sigma = 8.0$ wk, $c = 0.2$ wk^{-1}, $k = 5$. Assume time of initiation of significant growth is $t_i = 21$ wk. Calculate dimensionless time x and the growth quantifier Q from

$$x = \frac{t - \mu}{\sqrt{2}\sigma} + \frac{\sqrt{2}\sigma c}{2}$$

$$Q = \exp\left(\sqrt{2}\sigma c x_i\right)\left\{(1 - kx_i)[\text{erf } x - \text{erf } x_i]\right.$$
$$\left. - \frac{k}{\sqrt{\pi}}[\exp(-x^2) - \exp(-x_i^2)]\right\}$$

c. Calculate dry matter vs. time from the model equations (calibrated at $t = 28$ wk)

Nonirrigated: $\quad Y = AQ = \left(\dfrac{11.00}{2.49}\right)Q = 4.42Q$

Irrigated: $\quad Y = AQ = \left(\dfrac{10.50}{2.49}\right)Q = 4.22Q$

d. Plot the calculated values from (c) on (a).

e. Discuss the results. Does the model give a reasonable description of the data? Does the effect of irrigation appear to be adequately accounted for in the A parameter?

Growth response of corn to irrigation with subsoiling at Clayton, NC (1979)

Calendar time wk	Dry matter Mg ha^{-1}	
	Nonirrigated	Irrigated
15.6	—	—
20.9	0.24	0.23
21.9	0.39	0.47
23.3	2.10	2.32
24.7	4.18	3.81
26.9	10.20	9.56
28.6	11.82	11.13

Data adapted from Cassel (1988, table 9).

3.28 Cassel (1988) reported on response of corn to irrigation with subsoiling on Wagram loamy sand (loamy, siliceous, thermic Typic
Paleudult) at Clayton, NC. Data are given in the table below
for1980. Time is calculated from Jan. 1, with planting on April 19,
1980.

a. Plot dry matter Y vs. calendar time t for both treatments on linear
   graph paper.
b. Assume the following parameters for the growth model: $\mu = 26.0$
   wk, $\sqrt{2}\sigma = 8.0$ wk, $c = 0.2$ wk^{-1}, $k = 5$. Assume time of initiation
   of significant growth is $t_i = 20$ wk. Calculate dimensionless time x
   and the growth quantifier Q from

$$x = \frac{t - \mu}{\sqrt{2}\sigma} + \frac{\sqrt{2}\sigma c}{2}$$

$$Q = \exp\left(\sqrt{2}\sigma c x_i\right)\left\{(1 - kx_i)[\text{erf } x - \text{erf } x_i]\right.$$
$$\left. - \frac{k}{\sqrt{\pi}}[\exp(-x^2) - \exp(-x_i^2)]\right\}$$

c. Calculate dry matter vs. time from the model equations (calibrated
   at $t = 27$ wk)

$$\text{Nonirrigated:} \quad Y = AQ = \left(\frac{9.00}{2.36}\right)Q = 3.81Q$$

$$\text{Irrigated:} \quad Y = AQ = \left(\frac{10.50}{2.36}\right)Q = 4.45Q$$

d. Plot the calculated values from (c) on (a).
e. Discuss the results. Does the model give a reasonable description of
   the data? Does the effect of irrigation appear to be adequately
   accounted for in the A parameter?

Growth response of corn to irrigation with
subsoiling at Clayton, NC (1980)

Calendar time wk	Dry matter Mg ha^{-1}	
	Nonirrigated	Irrigated
15.6	—	—
18.3	0.01	0.01
20.0	0.15	0.17
21.4	0.68	0.84

23.4	3.21	4.01
25.6	7.11	7.88
27.0	8.39	10.22
29.0	11.79	14.72

Data adapted from Cassel (1988, table 10).

3.29 Cassel (1988) reported on response of corn to irrigation with subsoiling on Wagram loamy sand (loamy, siliceous, thermic Typic Paleudult) at Clayton, NC. Data are given in the table below for 1981. Time is calculated from Jan. 1, with planting on April 16, 1981.

a. Plot dry matter Y vs. calendar time t for both treatments on linear graph paper.

b. Assume the following parameters for the growth model: $\mu = 26.0$ wk, $\sqrt{2}\sigma = 8.0$ wk, $c = 0.2$ wk^{-1}, $k = 5$. Assume time of initiation of significant growth is $t_i = 21$ wk. Calculate dimensionless time x and the growth quantifier Q from

$$x = \frac{t - \mu}{\sqrt{2}\sigma} + \frac{\sqrt{2}\sigma c}{2}$$

$$Q = \exp\left(\sqrt{2}\sigma c x_i\right)\left\{(1 - kx_i)[\text{erf } x - \text{erf } x_i]\right.$$

$$\left. - \frac{k}{\sqrt{\pi}}[\exp(-x^2) - \exp(-x_i^2)]\right\}$$

c. Calculate dry matter vs. time from the model equations (calibrated at $t = 30$ wk)

Nonirrigated: $Y = AQ = \left(\dfrac{13.00}{3.05}\right)Q = 4.26Q$

Irrigated: $Y = AQ = \left(\dfrac{16.00}{3.05}\right)Q = 5.25Q$

d. Plot the calculated values from (c) on (a).

e. Discuss the results. Does the model give a reasonable description of the data? Does the effect of irrigation appear to be adequately accounted for in the A parameter?

Growth response of corn to irrigation with subsoiling at Clayton, NC (1981).

| Calendar time | Dry matter Mg ha^{-1} | |
| | Nonirrigated | Irrigated |
wk		
15.3	—	—
20.9	0.30	0.30
21.9	1.05	0.90
23.9	4.80	5.19
26.0	7.48	9.10
28.0	9.82	13.07
30.0	13.98	18.56
31.9	14.43	18.25

Data adapted from Cassel (1988, table 11).

3.30 Wright et al. (1988) studied response of corn to water availability (rainfall + irrigation) on Norfolk loamy fine sand (fine-loamy, kaolinitic, thermic Typic Paleudult) at Carrsville, VA. Applied N was 335 kg ha^{-1}.

   a. Plot grain yield Y vs. water availability W on linear graph paper for the four years and for nonirrigated, NI, and irrigated I treatments.

   b. Draw the curve of Y vs. W from the equation

$$Y = 16\{1 - \exp[-0.030(W - 15)]\}$$

   c. Since the model does not fit the data points for 1982, could this be related to excessive rainfall for that year? Speculate on the basis for this disagreement.

   d. Plot grain yield Y vs. evapotranspiration, ET, on linear graph paper for all years and both treatments.

   e. Draw the curve of Y vs. ET from the equation

$$Y = 16\{1 - \exp[-0.060(ET - 22)]\}$$

   f. Discuss the results.

Response of corn to water availability at Carrsville, VA

Year	Treatment	W cm	ET cm	Y Mg ha^{-1}
1980	NI	21.3	24.0	2.01
	I	51.4	44.2	10.50
1981	NI	35.4	37.7	6.41
	I	67.0	44.6	12.51
1982	NI	56.3	35.4	9.18
	I	78.4	36.9	10.56
1983	NI	35.8	34.3	7.95
	I	57.7	46.2	11.40

Data adapted from Wright et al. (1988, tables 3 and 4).

REFERENCES

Abramowitz, M., and I. A. Stegun. 1965. *Handbook of Mathematical Functions.* Dover Publications, New York.

Bar-Yosef, B., and U. Kafkafi. 1972. Rates of growth and nutrient uptake of irrigated corn as affected by N and P fertilization. *Soil Sci. Soc. Amer. Proc.* 36:931–936.

Beaty, E. R., J. D. Powell, R. H. Brown, and W. J. Ethredge. 1963. Effect of nitrogen rate and clipping frequency on yield of Pensacola bahiagrass. *Agron. J.* 55:3–4.

Blue, W. G. 1987. Response of pensacola bahiagrass (*Paspalum notatum* Flügge) to fertilizer nitrogen on an Entisol and a Spodosol in north Florida. *Soil and Crop Sci. Soc. Fla. Proc.* 47:135–139.

Burton, G. W., J. E. Jackson, and R. H. Hart. 1963. Effects of cutting frequency and nitrogen on yield, in vitro digestibility, and protein, fiber, and carotene content of coastal bermudagrass. *Agron. J.* 55:500–502.

Cassel, D. K. 1988. Clayton, North Carolina. In: *Scheduling Irrigation for Corn in the Southeast.* C. R. Camp and R. B. Campbell (eds). ARS-65. US Dept. of Agric. Washington, DC.

Christie, B. R., and A. R. McElroy. 1995. Orchardgrass. In: *Forages. Vol 1. An Introduction to Grassland Agriculture.* pp. 325–334. R. F. Barnes, D. A. Miller, and C. J. Nelson (eds). Iowa State University Press, Ames, IA.

Crowder, L. V., O. E. Sell, and E. M. Parker. 1955. The effect of clipping, nitrogen application, and weather on the productivity of fall sown oats, ryegrass and crimson clover. *Agron. J.* 47:51–54.

Day, J. L., and M. B. Parker. 1985. Fertilizer effects on crop removal of P and K in "coastal" bermudagrass forage. *Agron. J.* 77:110–114.

Deinum, B., and L. Sibma. 1980. Nitrate content of herbage in relation to nitrogen fertilization and management. In: The role of nitrogen in intensive grassland production. W. H. Prins and G. H. arnold (eds). *Proc. Int. Symp. Eur. Grassland Fed.*, Wageningen, The Netherlands.

Fulkerson, R. S. 1983. Research review of forage production. Crop Science Department. Guelph, Ontario, Canada.

Hammond, L. C., and D. Kirkham. 1949. Growth curves of soybeans and corn. *Agron. J.* 41:23–29.

Karlen, D. L., E. J. Sadler, and C. R. Camp. 1987. Dry matter, nitrogen, phosphorus, and potassium accumulation rates by corn on Norfolk loamy sand. *Agron. J.* 79:649–656.

Liu, F., C. C. Mitchell, J. W. Odom, D. T. Hill, and E. W. Rochester. 1997. Swine lagoon effluent disposal by overland flow: Effects on forage production and uptake of nitrogen and phosphorus. *Agron. J.* 89:900–904.

Mays, D. A., S. R. Wilkinson, and C. V. Cole. 1980. Phosphorus nutrition of forages. In: *The Role of Phosphorus in Agriculture.* F. E. Khasawneh, and E. J. Kamprath (eds.). American Society of Agronomy, Madison, WI.

Mutti, L.S.M. 1984. Dynamics of water and nitrogen stresses on corn growth, yield, and nutrient uptake. PhD Dissertation. University of Florida. Gainesville, FL.

Overman, A. R. 1984. Estimating crop growth rate with land treatment. *J. Env. Eng. Div., Amer. Soc. Civil Engr.* 110:1009–1012.

Overman, A. R. 1998. An expanded growth model for grasses. *Commun. Soil Sci. Plant Anal.* 29:67–85.

Overman, A. R. 1999. Model for accumulation of dry matter and plant nutrient elements by tobacco. *J. Plant Nutr.* 22:81–92.

Overman, A. R., and W. G. Blue. 1991. Estimation of dry matter production and nitrogen uptake by pensacola bahiagrass in Florida. *Florida Agric. Exp. Sta. Bull. 880.* Gainesville, FL.

Overman, A. R., and F. M. Rhoads. 1991. Estimation of dry matter production and nutrient removal by corn silage in Florida. *Florida Agric. Exp. Sta. Bull. 882.* University of Florida, Gainesville, FL.

Overman, A. R., and R. V. Scholtz III. 1999. Model for accumulation of dry matter and plant nutrients by corn. *Commun. Soil Sci. Plant Anal.* 30:2059–2081.

Overman, A. R., and R. L. Stanley. 1998. Bahiagrass response to applied nitrogen and harvest interval. *Commun. Soil Sci. Plant Anal.* 29:237–244.

Overman, A. R., E. A. Angley, and S. R. Wilkinson. 1988a. Empirical model of coastal bermudagrass production. *Trans. Amer. Soc. Agric. Engr.* 31:466–470.

Overman, A. R., E. A. Angley, and S. R. Wilkinson. 1988b. Evaluation of an empirical model of coastal bermudagrass production. *Agric. Sys.* 28:57–66.

Overman, A. R., E. A. Angley, and S. R. Wilkinson. 1989. A phenomenological model of coastal bermudagrass production. *Agric. Sys.* 29:137–148.

Overman, A. R., E. A. Angley; and S. R. Wilkinson. 1990a. Evaluation of a phenomenological model of coastal bermudagrass production. *Trans. Amer. Soc. Agric. Engr.* 33:443–450.

Overman, A. R., C. R. Neff, S. R. Wilkinson, and F. G. Martin. 1990b. Water, harvest interval, and applied nitrogen effects on forage yield of bermudagrass and bahiagrass. *Agron. J.* 82:1011–1016.

Overman, A. R., S. R. Wilkinson, and D. M. Wilson. 1994. An extended model of forage grass response to applied nitrogen. *Agron. J.* 86:617–620.

Prine, G. M., and G. W. Burton. 1956. The effect of nitrogen rate and clipping frequency upon the yield, protein content and certain morphological characteristics of coastal bermudagrass [*Cynodon dactylon* (L.) Pers.]. *Agron. J.* 48:296–301.

Prins, W. H., P. F. J. van Burg, and H. Wieling. 1980. The seasonal response of grassland to nitrogen at different intensities of nitrogen fertilization, with special reference to methods of response measurements. In: The role of nitrogen in intensive grassland production. W. H. Prins and G. H. Arnold (eds). *Proc. Int. Symp. Eur. Grassland Fed.*, Wageningen, The Netherlands.

Raper, Jr., C. D. and C. B. McCants. 1967. Nutrient accumulation in flue-cured tobacco. *Tobacco Sci.* 11:90.

Rhoads, F. M., and R. L. Stanley, Jr. 1979. Effect of population and fertility on nutrient uptake and yield on components of irrigated corn. *Soil and Crop Sci. Soc. Fla. Proc.* 38:78–81.

Sayre, J. D. 1948. Mineral accumulation in corn. *Plant Phys.* 23:267–281.

Soltanpour, P. N. 1969. Accumulation of dry matter and N, P, K, by Russet Burbank, Oromonte, and Red McClure potatoes. *Amer. Potato J.* 46:111–119.

Shih, S. F., and G. J. Gascho. 1980. Relationships among stalk, leaf area, and dry biomass of sugarcane. *Agron. J.* 72:309–313.

Staley, T. E., and H. D. Perry. 1995. Maize silage utilization of fertilizer and soil nitrogen on a hill-land ultisol relative to tillage method. *Agron. J.* 87:835–842.

Stanley, R. L. 1994. Response of "Tifton-9" Pensacola bahiagrass to harvest interval and nitrogen rate. *Soil Crop Sci. Soc. Fla. Proc.* 53:80–81.

Tolk, J. A., T. A. Howell, and S. R. Evett. 1998. Evapotranspiration and yield of corn grown on three high plains soils. *Agron. J.* 90:447–454.

Villanueva, M. R. 1974. Growth and nutritive quality of coastal bermudagrass [*Cynodon dactylon* (L.) Pers.] as influenced by nitrogen fertilization, defoliation and water management. PhD Dissertation. Texas A & M University. College Station, TX.

Wagger, M. G., and D. K. Cassel. 1993. Corn yield and water-use efficiency as affected by tillage and irrigation. *Soil Sci. Soc. Am. J.* 57:229–234.

Whitehead, D. C. 1995. *Grassland Nitrogen.* CAB International, Wallingford, UK.

Wright, F. S., N. L. Powell, and B. B. Ross. 1988. Suffolk, Virginia. In: *Scheduling Irrigation for Corn in the Southeast.* C. R. Camp and R. B. Campbell (eds). ARS-65. US Dept. of Agric., Washington, DC.

4

Mathematical Characteristics of Models

4.1 BACKGROUND

In this chapter attention is focused on mathematical details such as solution of differential equations and characteristics of some of the functions. A more rigorous mathematical foundation is also developed for the growth and seasonal models, including several theorems. Some readers may wish to skip certain details; these are provided to give a sound mathematical basis for the models.

4.2 PHENOMENOLOGICAL GROWTH MODEL

In Section 3.4 we introduced the phenomenological model. In this section this model is developed and details of integration of the resulting differential equation are given. The approach is a modification of the original by Overman et al. (1989). Since there is clearly a seasonal factor in the growth process, we postulate an environmental function E, defined by

$$E = \text{constant } \exp\left[-\left(\frac{t-\mu}{\sqrt{2}\sigma}\right)^2\right] \tag{4.1}$$

where t = calendar time since Jan. 1, wk; μ = time to mean of the environmental distribution, wk; σ = time spread of the environmental distribution, wk. The environmental factor may be solar input, for example. Then there is a factor related to the basic characteristics of the plant. Therefore, we define an intrinsic growth function given by the linear differential equation

$$\frac{dY'}{dt} = a + b\,(t - t_i)$$ [4.2]

where a, b = constants and t_i = reference time for the beginning of the growth interval, wk. Equation [4.2] describes growth rate dY'/dt under fixed environmental conditions without the seasonal effect. Finally, it is assumed that the net rate of growth dY/dt is the product of Eqs. [4.1] and [4.2], so that

$$\frac{dY}{dt} = \text{constant}\ [a + b(t - t_i)]\ \exp\left[-\left(\frac{t - \mu}{\sqrt{2}\sigma}\right)^2\right]$$ [4.3]

Now Eq. [4.3] contains two reference times, viz. t_i and μ. To facilitate integration of Eq. [4.3] it is convenient to introduce the dimensionless time variable x, defined by

$$x = \frac{t - \mu}{\sqrt{2}\sigma}$$ [4.4]

For a particular time t_i, Eq. [4.4] becomes

$$x_i = \frac{t_i - \mu}{\sqrt{2}\sigma}$$ [4.5]

It follows from Eqs. [4.4] and [4.5] that

$$\frac{t - t_i}{\sqrt{2}\sigma} = x - x_i$$ [4.6]

Equation [4.3] can now be rearranged to the form

$$\frac{dY}{dx} = \text{constant}\sqrt{2}\sigma\left[a + b\sqrt{2}\sigma(x - x_i)\right]\exp(-x^2)$$

$$= \text{constant}\ a\sqrt{2}\sigma\frac{\sqrt{\pi}}{2}\left[\frac{2}{\sqrt{\pi}} + \frac{2}{\sqrt{\pi}}\frac{\sqrt{2}\sigma b}{a}(x - x_i)\right]\exp(-x^2)$$

$$= A\left[\frac{2}{\sqrt{\pi}} + \frac{2}{\sqrt{\pi}}k(x - x_i)\right]\exp(-x^2)$$ [4.7]

where the parameters A and k are defined by

$$A = \frac{\text{constant}\ a\sqrt{2}\sigma\sqrt{\pi}}{2} = \text{yield factor}$$ [4.8]

$$k = \frac{\sqrt{2}\sigma b}{a} = \text{curvature factor} \tag{4.9}$$

The challenge now is to integrate Eq. [4.7]. Integration is taken from the reference time t_i to the general time $t > t_i$ so that the increment of dry matter accumulation ΔY_i is given by

$$\Delta Y_i = A \int_{x_i}^{x} \left[\frac{2}{\sqrt{\pi}} + \frac{2}{\sqrt{\pi}} k(x - x_i) \right] \exp(-x^2)\, dx \tag{4.10}$$

From mathematical tables (Abramowitz and Stegun, 1965) we find

$$\frac{2}{\sqrt{\pi}} \int_{x_i}^{x} \exp(-u^2)\, du = \text{erf } x - \text{erf } x_i \tag{4.11}$$

$$\int_{x_i}^{x} 2u \, \exp(-u^2)\, du = -\left[\exp(-x^2) - \exp(-x_i^2) \right] \tag{4.12}$$

where u is simply the variable of integration. It follows that the yield increment is given by

$$\Delta Y_i = A \left\{ (1 - kx_i)(\text{erf } x - \text{erf } x_i) - \frac{k}{\sqrt{\pi}} [\exp(-x^2) - \exp(-x_i^2)] \right\} \tag{4.13}$$

The model can be applied to perennial grasses which are harvested on a fixed time interval Δt by writing

$$\Delta Y_i = A \left\{ (1 - kx_i)(\text{erf } x_{i+1} - \text{erf } x_i) - \frac{k}{\sqrt{\pi}} [\exp(-x_{i+1}^2) - \exp(-x_i^2)] \right\} \tag{4.14}$$

where ΔY_i becomes the quantity of dry matter harvested at time t_{i+1} for growth interval i. It is convenient to define the growth quantifier ΔQ_i as

$$\Delta Q_i = (1 - kx_i)(\text{erf } x_{i+1} - \text{erf } x_i) - \frac{k}{\sqrt{\pi}} [\exp(-x_{i+1}^2) - \exp(-x_i^2)] \tag{4.15}$$

Then the cumulative dry matter for n harvests becomes

$$Y_n = \sum_{i=1}^{n} \Delta Y_i = A \sum_{i=1}^{n} \Delta Q_i \tag{4.16}$$

If Y_t is the total dry matter accumulation for all harvests, then normalized yield is given by

$$F = \frac{Y_n}{Y_t} \tag{4.17}$$

It was shown by Overman et al. (1990a) that a plot of F vs. t on probability paper produces a straight line. Overman et al. (1989) also showed that seasonal total dry matter follows a linear relationship to harvest interval

$$Y_t = \alpha + \beta \Delta t \qquad [4.18]$$

where α, β are regression coefficients.

We now prove a theorem to link seasonal total dry matter production Y_t to harvest interval Δt for a perennial grass. This amounts to deriving the seasonal growth quantifier from Eq. [4.16], where ΔQ_i can be considered piecewise continuous over the interval Δx.

Theorem. For the linear growth function with a fixed harvest interval Δt, seasonal total dry matter Y_t is related to harvest interval by

$$Y_t = 2A + \frac{kA}{\sqrt{2}\sigma} \Delta t \qquad [4.19]$$

Proof. Recall that the growth quantifier is defined by Eq. [4.15]. The summation process can therefore be broken into three parts. The first sum is described by

$$\sum_{i=1}^{n} (\text{erf } x_{i+1} - \text{erf } x_i) = \text{erf } x_2 - \text{erf } x_1 + \text{erf } x_3 - \text{erf } x_2 + \cdots$$

$$+ \text{erf } x_n - \text{erf } x_{i-1} = \text{erf } x_n - \text{erf } x_1 \qquad [4.20]$$

while the second sum is

$$\sum_{i=1}^{n} [\exp(-x_{i+1}^2) - \exp(-x_i^2)] = \exp(-x_2^2) - \exp(-x_1^2) + \exp(-x_3^2)$$

$$- \exp(-x_2^2) + \cdots + \exp(-x_n^2)$$
$$- \exp(-x_{n-1}^2)$$
$$= \exp(-x_n^2) - \exp(-x_1^2) \qquad [4.21]$$

It follows that over the entire domain $(-\infty, +\infty)$

$$\sum_{x_i=-\infty}^{x_i=+\infty} (\text{erf } x_{i+1} - \text{erf } x_i) = \text{erf }(\infty) - \text{erf }(-\infty) = 1 - (-1) = 2 \qquad [4.22]$$

and

$$\sum_{x_i=-\infty}^{x_i=+\infty} [\exp(-x_{i+1}^2) - \exp(-x_i^2)] = \exp(-\infty^2) - \exp(-\infty^2) = 0 - 0 = 0$$

[4.23]

The challenge is to prove that

$$\sum_{x_i=-\infty}^{x_i=+\infty} x_i(\text{erf } x_{i+1} - \text{erf } x_i) = -\Delta x$$

[4.24]

To do this, consider the graph shown in Fig. 4.1, where the auxiliary function, erf$'$, is defined by

$$\text{erf}' x = -\text{erf } x$$

[4.25]

Define the sum from left to right over the entire curve as

$$S = \sum^{\rightarrow} x_i(\text{erf } x_{i+1} - \text{erf } x_i)$$

[4.26]

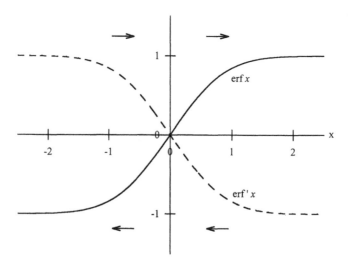

Figure 4.1 Characteristics of the error function.

The auxiliary sum is defined in a similar manner to obtain

$$S' = \sum \overrightarrow{x_i(\text{erf}'\, x_{i+1} - \text{erf}'\, x_i)} = \sum \overrightarrow{x_i(-\text{erf}\, x_{i+1} + \text{erf}\, x_i)}$$

$$= -\sum \overrightarrow{x_i(\text{erf}\, x_{i+1} - \text{erf}\, x_i)} = -S \qquad [4.27]$$

or

$$S + S' = 0 \qquad [4.28]$$

However, it should be noted that the summation process on the auxiliary function can be performed from right to left to obtain

$$S' = \sum \overleftarrow{x_{i+1}(\text{erf}'\, x_i - \text{erf}'\, x_{i+1})} = \sum \overleftarrow{x_{i+1}(-\text{erf}\, x_i + \text{erf}\, x_{i+1})}$$

$$= \sum \overleftarrow{x_{i+1}(\text{erf}\, x_{i+1} - \text{erf}\, x_i)} = -\sum \overrightarrow{x_i(\text{erf}\, x_{i+1} - \text{erf}\, x_i)} \qquad [4.29]$$

It follows from this last equation that

$$\sum (x_{i+1} - x_i)(\text{erf}\, x_{i+1} - \text{erf}\, x_i) = \sum x_{i+1}(\text{erf}\, x_{i+1} - \text{erf}\, x_i)$$

$$- \sum x_i(\text{erf}\, x_{i+1} - \text{erf}\, x_i)$$

$$= -2 \sum x_i\, (\text{erf}\, x_{i+1} - \text{erf}\, x_i) \qquad [4.30]$$

For a fixed harvest interval Δx, we have

$$\sum (x_{i+1} - x_i)(\text{erf}\, x_{i+1} - \text{erf}\, x_i) = \Delta x \sum (\text{erf}\, x_{i+1} - \text{erf}\, x_i) = 2\Delta x \qquad [4.31]$$

From Eqs. [4.30] and [4.31] we obtain finally

$$\sum x_i\, (\text{erf}\, x_{i+1} - \text{erf}\, x_i) = -\Delta x \qquad [4.32]$$

as required. It follows from these results that the seasonal growth quantifier Q_t is given by

$$Q_t = 2 + k\, \Delta x \qquad [4.33]$$

from which we obtain the linear relationship

$$Y_t = 2A + kA\, \Delta x = 2A + \frac{kA}{\sqrt{2}\sigma}\, \Delta t \qquad [4.34]$$

For regression purposes, Eq. [4.34] can be written as

$$Y_t = \alpha + \beta\, \Delta t \qquad \alpha, \beta = \text{constants} \qquad [4.35]$$

The question naturally arises as to the significance of the intrinsic growth function (Eq. [4.2]). Overman and Wilkinson (1989) used data from several studies (Burton et al., 1963; Day and Parker, 1985; Holt and Conrad, 1986; Prine and Burton, 1956) to support the idea that the a and b parameters relate to leaf and stem production, respectively, for bermudagrass. A broader interpretation would associate parameter a with the light-gathering component and parameter b with the structural component of the plant. It was shown from field data that the model only applies for growth intervals up to about 6 wk. A similar conclusion was reached with bahiagrass (Overman et al., 1989).

We now prove a more general theorem related to the phenomenological model, which will incorporate dry matter accumulation with time as well as seasonal total dry matter.

Theorem. For the linear growth function with a fixed harvest interval Δt, cumulative dry matter Y for n harvests follows the function

$$Y = A\left(2 + \frac{k}{\sqrt{2\sigma}}\Delta t\right)\left\{\frac{1}{2}\left[1 + \mathrm{erf}\left(\frac{t - \mu}{\sqrt{2\sigma}}\right)\right]\right\} \qquad [4.36]$$

Proof. The growth quantifier ΔQ_i for the linear intrinsic growth function is given by

$$\Delta Q_i = (1 - kx_i)(\mathrm{erf}\,x_{i+1} - \mathrm{erf}\,x_i) - \frac{k}{\sqrt{\pi}}\left[\exp(-x_{i+1}^2) - \exp(-x_i^2)\right] \qquad [4.37]$$

where k is the curvature factor. Dimensionless time x is related to calendar time t by

$$x = \frac{t - \mu}{\sqrt{2\sigma}} \qquad [4.38]$$

where μ = mean time of the environmental function, wk; and σ = spread of the environmental function, wk. The error function, erf, is defined by

$$\mathrm{erf}\,x = \frac{2}{\sqrt{\pi}} \int_0^x \exp(-u^2)\,du \qquad [4.39]$$

The cumulative growth quantifier for n growth intervals Q_n is given by

$$Q_n = \sum_{i=1}^n \Delta Q_i = \sum_{i=1}^n \left\{ (1 - kx_i)(\mathrm{erf}\,x_{i+1} - \mathrm{erf}\,x_i) - \frac{k}{\sqrt{\pi}}\left[\exp(-x_{i+1}^2) \right.\right.$$
$$\left.\left. - \exp(-x_i^2)\right] \right\} \qquad [4.40]$$

The first step is to shift x_i to the midpoint $x_{i+1/2}$ according to

$$x_{i+1/2} = x_i + \frac{\Delta x}{2} \tag{4.41}$$

for the fixed interval Δx. Equation [4.40] now becomes

$$
\begin{aligned}
Q_n &= \sum_{i=1}^{n}\left\{\left(1 + \frac{k\,\Delta x}{2} - kx_{i+1/2}\right)(\text{erf } x_{i+1} - \text{erf } x_i) - \frac{k}{\sqrt{\pi}}\,[\exp(-x_{i+1}^2) \right. \\
&\quad \left. - \exp(-x_i^2)]\right\} \\
&= \left(1 + \frac{k\Delta x}{2}\right)\sum_{i=1}^{n}(\text{erf } x_{i+1} - \text{erf } x_i) - k\sum_{i=1}^{n}x_{i+1/2}(\text{erf } x_{i+1} - \text{erf } x_i) \\
&\quad - \frac{k}{\sqrt{\pi}}\sum_{i=1}^{n}[\exp(-x_{i+1}^2) - \exp(-x_i^2)] \tag{4.42}
\end{aligned}
$$

Now we examine each of the three sums individually. The first sum can be expanded term by term to give

$$
\begin{aligned}
\sum_{i=1}^{n}(\text{erf } x_{i+1} - \text{erf } x_i) &= \text{erf } x_2 - \text{erf } x_1 + \text{erf } x_3 - \text{erf } x_2 + \cdots \\
&\quad + \text{erf} x_{n+1} - \text{erf} x_n = \text{erf} x_{n+1} - \text{erf} x_1 \tag{4.43}
\end{aligned}
$$

The cumulative sum over n harvests becomes

$$\sum_{x_i=-\infty}^{x_n}(\text{erf } x_{i+1} - \text{erf } x_i) = \text{erf } x_{n+1} - \text{erf}(-\infty) = \text{erf } x_{n+1} + 1 \tag{4.44}$$

It follows that the first term on the right side of Eq. [4.42] becomes

$$
\begin{aligned}
\left(1 + \frac{k\,\Delta x}{2}\right)\sum_{i=1}^{n}(\text{erf } x_{i+1} - \text{erf } x_i) &= \left(1 + \frac{k\,\Delta x}{2}\right)(1 + \text{erf } x_{n+1}) \\
&= (2 + k\,\Delta x)\left[\frac{1}{2}(1 + \text{erf } x_{n+1})\right] \tag{4.45}
\end{aligned}
$$

The next step is to examine the second term on the right side of Eq. [4.42]. We note that

$$x_{i+1/2}(\text{erf } x_{i+1} - \text{erf } x_i) = x_{i+1/2}\frac{2}{\sqrt{\pi}}\int_{x_i}^{x_{i+1}} \exp(-u^2)\, du$$

$$\approx \frac{2}{\sqrt{\pi}}\int_{x_i}^{x_{i+1}} u\exp(-u^2)\, du$$

$$= -\frac{1}{\sqrt{\pi}}[\exp(-x_{i+1}^2) - \exp(-x_i^2)] \qquad [4.46]$$

which leads to the sum

$$k\sum_{i=1}^{n} x_{i+1/2}(\text{erf } x_{i+1} - \text{erf } x_i) \approx -\frac{k}{\sqrt{\pi}}\sum_{i=1}^{n}[\exp(-x_{i+1}^2) - \exp(-x_i^2)]$$

$$[4.47]$$

It may be noted that this expression approximately cancels the third term on the right side of Eq. [4.42].

It remains to connect dry matter distribution to calendar time. Cumulative dry matter for n harvests Y_n is related to the cumulative growth quantifier by

$$Y_n = AQ_n = A(2 + k\,\Delta x)\left[\frac{1}{2}(1 + \text{erf } x_{n+1})\right] \qquad [4.48]$$

It was shown above that the seasonal growth quantifier Q_t is given by

$$Q_t = 2 + k\,\Delta x \qquad [4.49]$$

and seasonal yield Y_t by

$$Y_t = AQ_t = 2A + kA\,\Delta x \qquad [4.50]$$

Normalized yield fraction, $F = Y_n/Y_t$, is given by

$$F = \tfrac{1}{2}(1 + \text{erf } x) \qquad [4.51]$$

Since the dimensionless growth interval Δx is related to real growth interval Δt by

$$\Delta x = \frac{\Delta t}{\sqrt{2}\sigma} \qquad [4.52]$$

it follows that

$$F = \frac{1}{2}\left[1 + \text{erf}\left(\frac{t - \mu}{\sqrt{2}\sigma}\right)\right] \qquad [4.53]$$

and

$$Y_t = 2A + \frac{kA}{\sqrt{2}\sigma} \Delta t = \alpha + \beta \Delta t \qquad \alpha, \beta = \text{constants} \qquad [4.54]$$

Equations [4.53] and [4.54] establish the Gausssian distribution of F vs. t over the season and the linear dependence of Y_t on harvest interval Δt. This gives a basis for the empirical growth model, Eq. [3.1], published in 1984 as a simple beginning in my development of crop growth models (Overman, 1984).

4.3 EXPANDED GROWTH MODEL

The phenomenological model proved to be too restrictive for perennial grasses and did not apply to annuals such as corn. It became clear that the intrinsic growth function needed to be modified so that the growth rate did not continue unlimited, but would decrease to zero with time. This model has been presented by Overman (1998). We again assume the environmental function

$$E = \text{constant} \exp\left[-\left(\frac{t-\mu}{\sqrt{2}\sigma}\right)^2\right] \qquad [4.55]$$

Behavior of this function is shown in Fig. 4.2 for $\mu = 26$ wk and $\sqrt{2}\sigma = 8$ wk. In this case the intrinsic growth function is expanded to the linear-exponential form

$$\frac{dY'}{dt} = [a + b(t - t_i)] \exp[-c(t - t_i)] \qquad [4.56]$$

where $c =$ exponential parameter, wk^{-1}. Behavior of Eq. [4.56] is shown in Fig. 4.3 for $a = 1 = b$ and various values of c. The net growth rate is now assumed to be given by

$$\frac{dY}{dt} = \text{constant} [a + b(t - t_i)] \exp[-c(t - t_i)] \exp\left[-\left(\frac{t-\mu}{\sqrt{2}\sigma}\right)^2\right] \qquad [4.57]$$

The dimensionless time variable x' is defined as

$$x' = \frac{t - \mu}{\sqrt{2}\sigma} \qquad [4.58]$$

For a particular time t_i, Eq. [4.58] becomes

$$x_i' = \frac{t_i - \mu}{\sqrt{2}\sigma} \qquad [4.59]$$

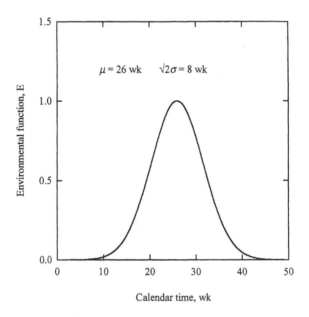

Figure 4.2 Characteristics of the Gaussian environmental function, Eq. [4.55].

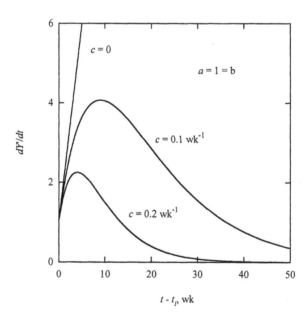

Figure 4.3 Characteristics of the linear-exponential intrinsic growth function, Eq. [4.56].

It follows from Eqs. [4.58] and [4.59] that

$$\frac{t - t_i}{\sqrt{2}\sigma} = x' - x_i'$$ [4.60]

The exponential terms can be simplified by completing the square to give

$$-x'^2 - c(t - t_i) = -x^2 + \sqrt{2}\sigma c x_i - \left(\frac{\sqrt{2}\sigma c}{2}\right)^2$$ [4.61]

where

$$x = x' + \frac{\sqrt{2}\sigma c}{2} = \frac{t - \mu}{\sqrt{2}\sigma} + \frac{\sqrt{2}\sigma c}{2}$$ [4.62]

and

$$x_i = x_i' + \frac{\sqrt{2}\sigma c}{2} = \frac{t_i - \mu}{\sqrt{2}\sigma} + \frac{\sqrt{2}\sigma c}{2}$$ [4.63]

It follows from Eqs. [4.62] and [4.63] that

$$\frac{t - t_i}{\sqrt{2}\sigma} = x - x_i$$ [4.64]

Equation [4.57] can now be written as

$$\frac{dY}{dx} = \text{constant } \sqrt{2}\sigma \, (a - \sqrt{2}\sigma b x_i$$

$$+ \sqrt{2}\sigma b x) \exp\left[-\left(\frac{\sqrt{2}\sigma c}{2}\right)^2 + \sqrt{2}\sigma c x_i\right] \exp(-x^2)$$ [4.65]

Equation [4.65] can be simplified by regrouping of terms

$$\frac{dY}{dx} = \text{constant } \sqrt{2}\sigma a \, \exp\left[-\left(\frac{\sqrt{2}\sigma c}{2}\right)^2\right]$$

$$\exp(\sqrt{2}\sigma c x_i)(1 - k x_i + k x) \exp(-x^2)$$ [4.66]

where the curvature factor is given by

$$k = \frac{\sqrt{2}\sigma b}{a}$$ [4.67]

It follows that incremental growth for the ith interval is given by

$$\Delta Y_i = A \, \Delta Q_i$$ [4.68]

where the yield factor A is defined by

$$A = \text{constant } \sqrt{2}\sigma a \left(\frac{\sqrt{\pi}}{2}\right) \exp\left[-\left(\frac{\sqrt{2}\sigma c}{2}\right)^2\right] \qquad [4.69]$$

and the growth quantifier is now defined by

$$\Delta Q_i = \exp(\sqrt{2}\sigma c x_i)$$

$$\left\{(1 - kx_i)(\text{erf } x - \text{erf } x_i) - \frac{k}{\sqrt{\pi}} [\exp(-x^2) - \exp(-x_i^2)]\right\} \qquad [4.70]$$

Note that for $c = 0$ this model reduces to the simple phenomenological model in Section 4.2.

Some of the characteristics of the expanded model have been discussed by Overman (1998). It was shown for a perennial grass harvested on a fixed interval Δt that normalized yield vs. time follows a straight line on probability paper. It was further shown that seasonal yield, Y_t, relates to harvest interval through the linear-exponential equation

$$Y_t = (\alpha + \beta \Delta t) \exp(-\gamma \Delta t) \qquad [4.71]$$

where α, β, γ are regression coefficients. Both of these consequences agree with field data for bermudagrass and for bahiagrass (Overman and Stanley, 1998). At this point Eq. [4.71] has not been established on rigorous grounds. A proof could possibly be based on the Gauss-Weierstrass transformation (Zemanian, 1987). There surely is linkage between Eqs. [4.56] and [4.71]. Further work is needed to identify the significance of the exponential term in Eq. [4.56]. The term occurs as a multiplier in the intrinsic growth function, and is central to the success of this model in describing both perennial grasses and annuals such as corn. It could be called an "aging factor."

We now prove a conditional theorem for the linear-exponential growth function, which applies to the case where ΔQ_i is assumed to follow a Gaussian distribution for fixed harvest interval Δt.

Theorem (Conditional). For the expanded growth function, seasonal total dry matter Y_t is related to a fixed harvest interval Δt by

$$Y_t = \left(2A' + \frac{kA'}{\sqrt{2}\sigma}\Delta t\right) \exp\left(-\frac{c}{2}\Delta t\right) \qquad [4.72]$$

if cumulative dry matter follows a Gaussian distribution.

Proof. The first step is to transpose Eq. [4.70] to the form

$$\Delta Q_i \exp(-\sqrt{2}\sigma c x_i) = (1 - k x_i)(\text{erf } x_{i+1} - \text{erf } x_i)$$
$$- \frac{k}{\sqrt{\pi}} [\exp(-x_{i+1}^2) - \exp(x_i^2)] \qquad [4.73]$$

The next step is to sum over the entire season

$$\sum_{x_i=-\infty}^{x_i=+\infty} \Delta Q_i \exp(-\sqrt{2}\sigma c x_i) = \sum_{x_i=-\infty}^{x_i=+\infty} \left\{ (1 - k x_i)(\text{erf } x_{i+1} - \text{erf } x_i) \right.$$
$$\left. - \frac{k}{\sqrt{\pi}} [\exp(-x_{i+1}^2) - \exp(x_i^2)] \right\} \qquad [4.74]$$

We have already shown that the right side of Eq. [4.74] is given by

$$\sum_{x_i=-\infty}^{x_i=+\infty} \left\{ (1 - k x_i)(\text{erf } x_{i+1} - \text{erf } x_i) - \frac{k}{\sqrt{\pi}} [\exp(-x_{i+1}^2) - \exp(-x_i^2)] \right\}$$
$$= 2 + k \, \Delta x \qquad [4.75]$$

The left side of Eq. [4.74] can be adjusted to the midpoint by adding $\Delta x/2$ so that

$$\Delta Q_i \exp(-\sqrt{2}\sigma c x_i) = \Delta Q_i \exp\left[-\sqrt{2}\sigma c \left(x_i + \frac{\Delta x}{2} \right) + \frac{\sqrt{2}\sigma c \Delta x}{2} \right]$$
$$= \exp\left(\frac{\sqrt{2}\sigma c \Delta x}{2} \right) \Delta Q_i \exp(-\sqrt{2}\sigma x_{i+1/2}) \qquad [4.76]$$

Now we assume that ΔQ_i follows a Gaussian distribution over the season, i.e.

$$\Delta Q_i = \exp\left[-\left(\frac{t_{i+1/2} - \mu}{\sqrt{2}\sigma} \right)^2 \right] \qquad [4.77]$$

aside from an arbitrary constant and where t = calendar time since Jan. 1, wk; μ = mean of the environmental function, wk; and σ = spread of the environmental function, wk. The dimensionless time variable is related to calendar time by

$$x = \frac{t - \mu}{\sqrt{2}\sigma} + \frac{\sqrt{2}\sigma c}{2} \qquad [4.78]$$

It follows that

$$\left(\frac{t_{i+1/2} - \mu}{\sqrt{2}\sigma}\right)^2 = \left(x_{i+1/2} - \frac{\sqrt{2}\sigma c}{2}\right)^2 = x_{i+1/2}^2 - \sqrt{2}\sigma c x_{i+1/2} + \left(\frac{\sqrt{2}\sigma c}{2}\right)^2$$

[4.79]

and that

$$-\left(\frac{t_{i+1/2} - \mu}{\sqrt{2}\sigma}\right)^2 - \sqrt{2}\sigma c x_{i+1/2} = -x_{i+1/2}^2 + \sqrt{2}\sigma c x_{i+1/2} - \left(\frac{\sqrt{2}\sigma c}{2}\right)^2$$

$$- \sqrt{2}\sigma c x_{i+1/2} = -x_{i+1/2}^2 - \left(\frac{\sqrt{2}\sigma c}{2}\right)^2$$

[4.80]

According to Eq. [4.80] if ΔQ_i follows a Gaussian distribution, then the shifted function Eq. [4.76] also follows a Gaussian distribution. The left side of Eq. [4.74] can now be written as

$$\exp\left(\frac{\sqrt{2}\sigma c \Delta x}{2}\right) \sum_{x_i=-\infty}^{x_i=+\infty} \Delta Q_i \exp(-\sqrt{2}\sigma c x_{i+1/2})$$

$$= \exp\left(\frac{\sqrt{2}\sigma c \Delta x}{2}\right) \exp\left[-\left(\frac{\sqrt{2}\sigma c}{2}\right)^2\right] \sum_{x_i=-\infty}^{x_i=+\infty} \exp(-x_{i+1/2}^2)$$

$$= \exp\left(\frac{\sqrt{2}\sigma c \Delta x}{2}\right) \exp\left[-\left(\frac{\sqrt{2}\sigma c}{2}\right)^2\right] Q_t$$

[4.81]

The last step follows from the fact that

$$Q_t = \sum_{x_i=-\infty}^{x_i=+\infty} \Delta Q_i = \sum_{x_i=-\infty}^{x_i=+\infty} \exp\left[-\left(\frac{t_{i+1/2} - \mu}{\sqrt{2}\sigma}\right)^2\right] = \sum_{x_i=-\infty}^{x_i=+\infty} \exp(-x_{i+1/2}^2)$$

[4.82]

since the total sum over t and x must be the same. Combination of Eqs. [4.75] and [4.82] leads to

$$Y_t = A \exp\left[\left(\frac{\sqrt{2}\sigma c}{2}\right)^2\right] \exp\left[-\left(\frac{\sqrt{2}\sigma c}{2}\Delta x\right)\right](2 + k \Delta x)$$

[4.83]

This procedure can be simplified somewhat. In the expanded growth model we defined the yield factor A as

$$A = \text{constant } \sqrt{2}\sigma a\left(\frac{\sqrt{\pi}}{2}\right) \exp\left[-\left(\frac{\sqrt{2}\sigma c}{2}\right)^2\right] \qquad [4.84]$$

Now define a modified yield factor A' as

$$A' = \text{constant } \sqrt{2}\sigma a\left(\frac{\sqrt{\pi}}{2}\right) \qquad [4.85]$$

which is the same as for the linear growth model, and move the exponential term to a modified seasonal growth quantifier Q'_t, defined by

$$Q'_t = Q_t \exp\left[-\left(\frac{\sqrt{2}\sigma c}{2}\right)^2\right] = (2 + k\Delta x) \exp\left(-\frac{\sqrt{2}\sigma c}{2}\Delta x\right)$$

$$= \left(2 + \frac{k}{\sqrt{2}\sigma}\Delta t\right)\exp\left(-\frac{c}{2}\Delta t\right) \qquad [4.86]$$

Equation [4.86] establishes the simple linear-exponential relationship of the seasonal growth quantifier to harvest interval. The linear term is the same as for the linear growth model. The linear-exponential growth model simply adds the exponential term to the seasonal growth quantifier. Finally, we link seasonal total yield to harvest interval by

$$Y_t = A'Q'_t = \left(2A' + \frac{kA'}{\sqrt{2}\sigma}\Delta t\right)\exp\left(-\frac{c}{2}\Delta t\right) \qquad [4.87]$$

which completes proof of the conditional theorem.

The next step is to prove a general theorem related to the expanded growth model, which will relate dry matter accumulation to time and harvest interval. The conditional theorem above serves as an intermediate step to this more general form, and is included to illustrate how such problems are often solved in a step-wise manner.

Theorem (General). For the linear-exponential growth function with a fixed harvest interval Δt cumulative dry matter Y for n harvests follows the function

$$Y = A'\left(2 + \frac{k}{\sqrt{2}\sigma}\Delta t\right)\exp\left(-\frac{c}{2}\Delta t\right)\left\{\frac{1}{2}\left[1 + \text{erf}\left(\frac{t - \mu}{\sqrt{2}\sigma}\right)\right]\right\} \qquad [4.88]$$

Proof. The growth quantifier ΔQ_i for the linear-exponential intrinsic growth function is given by

$$\Delta Q_i = \exp(\sqrt{2}\sigma c x_i)$$

$$\left\{ (1 - kx_i)(\operatorname{erf} x_{i+1} - \operatorname{erf} x_i) - \frac{k}{\sqrt{\pi}} [\exp(-x_{i+1}^2) - \exp(-x_i^2)] \right\} \quad [4.89]$$

where k is the curvature factor and c is the aging factor, wk^{-1}. Dimensionless time x for this model is related to calendar time t by

$$x = \frac{t - \mu}{\sqrt{2}\sigma} + \frac{\sqrt{2}\sigma c}{2} \quad [4.90]$$

where μ = mean time of the environmental function, wk; and σ = spread of the environmental function, wk. The error function, erf, is defined by

$$\operatorname{erf} x = \frac{2}{\sqrt{\pi}} \int_0^x \exp(-u^2)\, du \quad [4.91]$$

The cumulative growth quantifier for n growth intervals Q_n is given by

$$Q_n = \sum_{i=1}^{n} \Delta Q_i = \sum_{i=1}^{n} \exp(\sqrt{2}\sigma c x_i)\Big\{ (1 - kx_i)(\operatorname{erf} x_{i+1} - \operatorname{erf} x_i)$$

$$- \frac{k}{\sqrt{\pi}} [\exp(-x_{i+1}^2) - \exp(-x_i^2)] \Big\} \quad [4.92]$$

The first step is to shift x_i to the midpoint $x_{i+1/2}$ according to

$$x_{i+1/2} = x_i + \frac{\Delta x}{2} \quad [4.93]$$

for the fixed interval Δx. Equation [4.92] now becomes

$$Q_n = \sum_{i=1}^{n} \exp\left(\sqrt{2}\sigma c x_{i+1/2} - \sqrt{2}\sigma c \frac{\Delta x}{2} \right)$$

$$\left\{ (1 + \frac{k\Delta x}{2} - kx_{i+1/2})(\operatorname{erf} x_{i+1} - \operatorname{erf} x_i) - \frac{k}{\sqrt{\pi}} [\exp(-x_{i+1}^2) - \exp(-x_i^2)] \right\}$$

$$= \left(1 + \frac{k\Delta x}{2} \right) \exp\left(-\sqrt{2}\sigma c \frac{\Delta x}{2} \right) \sum_{i=1}^{n} (\operatorname{erf} x_{i+1} - \operatorname{erf} x_i) \exp(\sqrt{2}\sigma c x_{i+1/2})$$

$$- k \exp\left(-\sqrt{2}\sigma c \frac{\Delta x}{2} \right) \sum_{i=1}^{n} x_{i+1/2}(\operatorname{erf} x_{i+1} - \operatorname{erf} x_i) \exp(\sqrt{2}\sigma c x_{i+1/2})$$

$$- \frac{k}{\sqrt{\pi}} \exp\left(-\sqrt{2}\sigma c \frac{\Delta x}{2} \right) \sum_{i=1}^{n} [\exp(-x_{i+1}^2) - \exp(-x_i^2)] \exp(\sqrt{2}\sigma c x_{i+1/2})$$

$$[4.94]$$

Now we examine each of the three sums individually. From the first sum we note that

$$(\text{erf } x_{i+1} - \text{erf } x_i)\exp(\sqrt{2}\sigma c x_{i+1/2})$$

$$= \exp(\sqrt{2}\sigma c x_{i+1/2})\frac{2}{\sqrt{\pi}}\int_{x_i}^{x_{i+1}}\exp(-u^2)\,du$$

$$\approx \frac{2}{\sqrt{\pi}}\int_{x_i}^{x_{i+1}}\exp(-u^2 + \sqrt{2}\sigma c u)\,du \tag{4.95}$$

Equation [4.95] can be simplified by completing the square in the exponent

$$-u^2 + \sqrt{2}\sigma c u = -u^2 + \sqrt{2}\sigma c u - \left(\frac{\sqrt{2}\sigma c}{2}\right)^2 + \left(\frac{\sqrt{2}\sigma c}{2}\right)^2$$

$$= -\left(u - \frac{\sqrt{2}\sigma c}{2}\right)^2 + \left(\frac{\sqrt{2}\sigma c}{2}\right)^2 \tag{4.96}$$

Substitution of Eq. [4.96] into Eq. [4.95] leads to

$$\frac{2}{\sqrt{\pi}}\int_{x_i}^{x_{i+1}}\exp(-u^2 + \sqrt{2}\sigma c u)\,du = \exp\left[\left(\frac{\sqrt{2}\sigma c}{2}\right)^2\right]\left[\text{erf}\left(x_{i+1} - \frac{\sqrt{2}\sigma c}{2}\right)\right.$$

$$\left. - \text{erf}\left(x_i - \frac{\sqrt{2}\sigma c}{2}\right)\right] \tag{4.97}$$

From Eq. [4.90] it follows that

$$x - \frac{\sqrt{2}\sigma c}{2} = \frac{t - \mu}{\sqrt{2}\sigma} = x' \tag{4.98}$$

With this simplification we can write

$$\sum_{i=1}^{n}(\text{erf } x_{i+1} - \text{erf } x_i)\exp(\sqrt{2}\sigma c x_{i+1/2})$$

$$\approx \exp\left[\left(\frac{\sqrt{2}\sigma c}{2}\right)^2\right]\sum_{i=1}^{n}(\text{erf } x'_{i+1} - \text{erf } x'_i) \tag{4.99}$$

The sum is examined term by term to give

$$\sum_{i=1}^{n}(\text{erf } x'_{i+1} - \text{erf } x'_i) = \text{erf } x'_2 - \text{erf } x'_1 + \text{erf } x'_3 - \text{erf } x'_2 + \cdots$$

$$+ \text{erf } x'_{n+1} - \text{erf } x'_n = \text{erf } x'_{n+1} - \text{erf } x'_1 \tag{4.100}$$

The cumulative sum over n harvests becomes

$$\sum_{x_i=-\infty}^{x_n} (\text{erf } x'_{i+1} - \text{erf } x'_i) = \text{erf } x'_{n+1} - \text{erf}(-\infty) = 1 + \text{erf } x'_{n+1} \qquad [4.101]$$

It follows that the first term on the right side of Eq. [4.94] becomes

$$\left(1 + \frac{k\,\Delta x}{2}\right)\exp\left(-\sqrt{2}\sigma c\,\frac{\Delta x}{2}\right)\sum_{i=1}^{n}(\text{erf } x_{i+1} - \text{erf } x_i)\exp(\sqrt{2}\sigma c x_{i+1/2})$$

$$\approx \exp\left[\left(\frac{\sqrt{2}\sigma c}{2}\right)^2\right]\left(1 + \frac{k\,\Delta x}{2}\right)\exp\left(-\sqrt{2}\sigma c\,\frac{\Delta x}{2}\right)[1 + \text{erf } x'_{n+1}]$$

$$= \exp\left[\left(\frac{\sqrt{2}\sigma c}{2}\right)^2\right](2 + k\,\Delta x)\exp\left(-\sqrt{2}\sigma c\,\frac{\Delta x}{2}\right)\left[\frac{1}{2}(1 + \text{erf } x'_{n+1})\right]$$

$$[4.102]$$

Next we focus on the second sum in Eq. [4.94]. It is convenient to use the approximation

$$x_{i+1/2}(\text{erf } x_{i+1} - \text{erf } x_i) = x_{i+1/2}\frac{2}{\sqrt{\pi}}\int_{x_i}^{x_{i+1}}\exp(-u^2)\,du$$

$$\approx \frac{2}{\sqrt{\pi}}\int_{x_i}^{x_{i+1}} u\exp(-u^2)\,du$$

$$= -\frac{1}{\sqrt{\pi}}[\exp(-x_{i+1}^2) - \exp(-x_i^2)] \qquad [4.103]$$

It follows from this result that the second and third terms in Eq. [4.94] cancel and leave only the first term. Recalling Eq. [4.98] and that

$$\Delta x = \frac{\Delta t}{\sqrt{2}\sigma} \qquad [4.104]$$

we obtain

$$Q_n \approx \exp\left[\left(\frac{\sqrt{2}\sigma c}{2}\right)^2\right]\left(2 + \frac{k}{\sqrt{2}\sigma}\Delta t\right)\exp\left(-\frac{c}{2}\Delta t\right)\left\{\frac{1}{2}\left[1 + \text{erf}\left(\frac{t - \mu}{\sqrt{2}\sigma}\right)\right]\right\}$$

$$[4.105]$$

For the expanded growth model we defined the yield factor A by Eq. [4.69]:

$$A = \text{constant } \sqrt{2}\sigma a\left(\frac{\sqrt{\pi}}{2}\right)\exp\left[-\left(\frac{\sqrt{2}\sigma c}{2}\right)^2\right] \qquad [4.106]$$

It is now convenient to define a modified yield factor A' as

$$A' = \text{constant } \sqrt{2}\sigma a\left(\frac{\sqrt{\pi}}{2}\right)$$ [4.107]

and move the exponential factor to the growth quantifier and define a modified growth quantifier Q'_n as

$$Q'_n \approx \left\{\left(2 + \frac{k}{\sqrt{2}\sigma}\,\Delta t\right)\exp\left(-\frac{c}{2}\,\Delta t\right)\right\}\left\{\frac{1}{2}\left[1 + \text{erf}\left(\frac{t-\mu}{\sqrt{2}\sigma}\right)\right]\right\}$$ [4.108]

Normalized yield fraction F is defined by

$$F = \frac{1}{2}\left[1 + \text{erf}\left(\frac{t-\mu}{\sqrt{2}\sigma}\right)\right]$$ [4.109]

and seasonal growth quantifier by

$$Q'_t \approx \left(2 + \frac{k}{\sqrt{2}\sigma}\,\Delta t\right)\exp\left(-\frac{c}{2}\,\Delta t\right)$$ [4.110]

Now seasonal total yield is given by

$$Y_t = A'Q'_t = \left(2A' + \frac{kA'}{\sqrt{2}\sigma}\,\Delta t\right)\exp\left(-\frac{c}{2}\,\Delta t\right)$$
$$= (\alpha + \beta\,\Delta t)\exp(-\gamma\,\Delta t) \qquad \alpha, \beta, \gamma = \text{constants}$$ [4.111]

Equation [4.109] establishes the Gaussian distribution of F vs. t over the season, while Eq. [4.111] shows the linear-exponential relationship between Y_t and harvest interval Δt. Equations [4.109] and [4.111] form operational definitions which can be used in data analysis. This theorem gives further basis for the empirical model, Eq. [3.1].

Two characteristics of the general theorem should be noted. First, the Gaussian component of the expanded growth model transforms to a Gaussian distribution of dry matter over the season. Second, the linear-exponential intrinsic growth function transforms to a linear-exponential dependence of seasonal total dry matter on harvest interval. Both of these characteristics have been noted in field data for perennial grasses.

4.3.1 Alternative Development

We now present an alternative approach to the expanded growth model for perennial grass harvested on a fixed time interval. A central feature of the mathematical approach is the Dirac delta function $\delta(x)$ introduced by Dirac (1958, p. 58) in 1927 and used extensively in mathematical physics (Bracewell, 2000, chap. 5), and which is defined by

$$\delta(x) = 0 \qquad x \neq 0$$

$$\int_{-\infty}^{+\infty} \delta(x)\, dx = 1 \qquad\qquad [4.112]$$

In other words, the delta function represents a pulse of unit area at $x = 0$ in this case. The pulse can be moved to $x = a$ by the definition

$$\delta(x - a) = 0 \qquad x \neq a$$

$$\int_{-\infty}^{+\infty} \delta(x - a)\, dx = 1 \qquad\qquad [4.113]$$

It can be shown for a function $f(x)$ that

$$\int_{-\infty}^{+\infty} \delta(x - a)f(x)\, dx = f(a) \qquad\qquad [4.114]$$

This is called the *sifting property*, which simply gives the value of the function at $x = a$ and is zero at all other points. An even more important characteristic for our purpose is the multiple delta function designated by the shah symbol, $III(x)$, and defined by

$$III(x) = \sum_{a_i=-\infty}^{a_i=+\infty} \delta(x - a_i) \qquad\qquad [4.115]$$

where pulses occur at discrete points a_i on a uniform spacing Δx. We now obtain for the function $f(x)$

$$\int_{-\infty}^{+\infty} III(x)f(x)\, dx = \int_{-\infty}^{+\infty} \sum_{a_i=-\infty}^{a_i=+\infty} \delta(x - a_i)f(x)\, dx$$

$$= \sum_{a_i=-\infty}^{a_i=+\infty} \int_{-\infty}^{+\infty} \delta(x - a_i)f(x)\, dx = \sum_{a_i=-\infty}^{a_i=+\infty} f(a_i) \qquad [4.116]$$

This is called the *sampling property*, which gives the values of the function at multiple points a_i and is zero at all other points.

To apply the sampling property to the expanded growth model, we recall the differential equation (Eq. [4.66])

$$\frac{dY}{dx} = \text{constant } \sqrt{2}\sigma a \exp\left[-\left(\frac{\sqrt{2}\sigma c}{2}\right)^2\right]$$

$$\exp(\sqrt{2}\sigma c x_i)(1 - kx_i + kx)\exp(-x^2) \qquad\qquad [4.117]$$

It is convenient at this point to define

$$f(x) = \exp\left[-\left(\frac{\sqrt{2}\sigma c}{2}\right)^2\right]\exp(\sqrt{2}\sigma c x_i)(1 - kx_i + kx)\exp(-x^2) \qquad [4.118]$$

For this analysis it is appropriate to evaluate the function at the midpoint of each interval, $x = x_{i+1/2}$, so that

$$1 + k(x - x_i) = 1 + k(x_{i+1/2} - x_i) = 1 + k\frac{\Delta x}{2} \qquad [4.119]$$

and

$$
\begin{aligned}
-x^2 + \sqrt{2}\sigma c x_i &= -x_{i+1/2}^2 + \sqrt{2}\sigma c\left(x_{i+1/2} - \frac{\Delta x}{2}\right) \\
&= -x_{i+1/2}^2 + \sqrt{2}\sigma c x_{i+1/2} - \left(\frac{\sqrt{2}\sigma c}{2}\right)^2 + \left(\frac{\sqrt{2}\sigma c}{2}\right)^2 \\
&\quad - \sqrt{2}\sigma c\frac{\Delta x}{2} \\
&= -\left(x_{i+1/2} - \frac{\sqrt{2}\sigma c}{2}\right)^2 + \left(\frac{\sqrt{2}\sigma c}{2}\right)^2 - \sqrt{2}\sigma c\frac{\Delta x}{2} \qquad [4.120]
\end{aligned}
$$

It follows that Eq. [4.118] becomes

$$f(x_{i+1/2}) = \left(1 + k\frac{\Delta x}{2}\right)\exp\left(-\sqrt{2}\sigma c\frac{\Delta x}{2}\right)\exp\left[-\left(x_{i+1/2} - \frac{\sqrt{2}\sigma c}{2}\right)^2\right]$$

$$[4.121]$$

Recalling the relationship between x and t,

$$x = \frac{t - \mu}{\sqrt{2}\sigma} + \frac{\sqrt{2}\sigma c}{2} \qquad [4.122]$$

Eq. [4.121] can be written as

$$f(t_{i+1/2}) = \left(1 + \frac{k}{\sqrt{2}\sigma}\frac{\Delta t}{2}\right)\exp\left(-c\frac{\Delta t}{2}\right)\exp\left[-\left(\frac{t_{i+1/2} - \mu}{\sqrt{2}\sigma}\right)^2\right] \qquad [4.123]$$

Now the sum of the Gaussian terms over the entire spectrum is

$$\sum_{t_{i+1/2}=-\infty}^{t_{i+1/2}=+\infty}\exp\left[-\left(\frac{t_{i+1/2} - \mu}{\sqrt{2}\sigma}\right)^2\right] = \sqrt{\pi} \qquad [4.124]$$

for $\Delta x \leq 1$. For $\Delta x > 1$ the sum in Eq. [4.124] is less than $\sqrt{\pi}$. Total dry matter for the season Y_t is given by

$$Y_t = \text{constant } \sqrt{2}\sigma a \sum_{t_{i+1/2}=-\infty}^{t_{i+1/2}=+\infty} f(t_{i+1/2})$$

$$= \text{constant } \sqrt{2}\sigma a \sqrt{\pi} \left(1 + \frac{k}{\sqrt{2}\sigma} \frac{\Delta t}{2}\right) \exp\left(-c \frac{\Delta t}{2}\right)$$

$$= \text{constant } \sqrt{2}\sigma a \frac{\sqrt{\pi}}{2} \left(2 + \frac{k}{\sqrt{2}\sigma} \Delta t\right) \exp\left(-\frac{c}{2} \Delta t\right)$$

$$= A'\left(2 + \frac{k}{\sqrt{2}\sigma} \Delta t\right) \exp\left(-\frac{c}{2} \Delta t\right) \qquad [4.125]$$

where the modified yield factor A' is given by

$$A' = \text{constant } \sqrt{2}\sigma a \left(\frac{\sqrt{\pi}}{2}\right) \qquad [4.126]$$

Equation [4.125] is the same as obtained previously and shows again the linear-exponential dependence of total seasonal yield Y_t on harvest interval Δt.

4.4 RATIONAL BASIS FOR LOGISTIC MODEL

Previous work has shown the usefulness of the logistic model for relating forage yield response to applied nutrients (Overman et al., 1990b, 1990c; Overman et al., 1991; Overman and Evers, 1992; Overman and Wilkinson, 1992). The model has been extended to couple dry matter and plant N uptake (Overman et al., 1994). A consequence of the extended model is that dry matter is related to plant N uptake through a hyperbolic equation. In this section a rational basis is presented for the logistic equation, as previously discussed by Overman (1995a).

At very low levels of applied N, seasonal dry matter follows geometric response to applied N:

$$\frac{dY}{dN} = kY \quad \text{with } Y = Y_0 \quad \text{at } N = 0 \qquad [4.127]$$

where $k = $ N response coefficient. This model assumes that incremental response relates to the quantity of dry matter present. Equation [4.127] can be rearranged by separating the variables

$$\frac{dY}{Y} = k \, dN \qquad [4.128]$$

Integration of Eq. [4.128] leads to

$$Y = Y_0 \exp(kN) \qquad [4.129]$$

which describes positive response to applied N for $k > 0$ and $Y_0 > 0$. For regression purposes, Eq. [4.129] can be viewed as the two parameter model

$$Y = A \exp(cN) \qquad [4.130]$$

Now the model can be linearized by taking logarithms so that

$$\ln Y = \ln A + cN = b + cN \qquad [4.131]$$

where $b - \ln A$. Clearly, Eq. [4.129] can only have limited validity.

At higher levels of applied N, response appears to follow a different exponential equation

$$\frac{dY}{dN} = k(Y_m - Y) \qquad \text{with } Y = Y_0 \text{ at } N = 0 \qquad [4.132]$$

with Y_m = maximum seasonal dry matter at high applied N. This is in fact the famous Mitscherlich equation (Mengel and Kirkby, 1987; Russell, 1950). It assumes that incremental response to applied N is proportional to the unfilled capacity $Y_m - Y$ of the system. Separation of variables in Eq. [4.132] leads to

$$\frac{dY}{Y_m - Y} = k \, dN \qquad [4.133]$$

Integration of Eq. [4.133] gives

$$Y_m - Y = (Y_m - Y_0) \exp(-kN) \qquad [4.134]$$

or to the alternative form

$$Y = Y_0 + (Y_m - Y_0)[1 - \exp(-kN)] \qquad [4.135]$$

This model exhibits asymptotic approach to maximum yield for $k > 0$. For regression purposes this represents a three-parameter model

$$Y = A - B \exp(-cN) \qquad [4.136]$$

where A, B, c are regression coefficients to be evaluated from data. The linearized form of Eq. [4.136] is

$$\ln(A - Y) = \ln B - cN = b - cN \qquad [4.137]$$

where $b = \ln B$.

The exponential and Mitscherlich models present a dilemma. At high applied N, Eq. [4.130] increases without limit. At low applied N, Eq. [4.135] can go negative, which is obviously unacceptable. Our challenge is to find a compromise model which converges at both low and high applied

N. We follow the same procedure as Planck in guessing the correct equation for black body radiation (Longair, 1984, p. 205). At low applied N and low Y we found

$$\text{Lower N:} \quad \frac{dY}{dN} \propto Y \tag{4.138}$$

While at higher applied N:

$$\text{Higher N:} \quad \frac{dY}{dN} \propto Y_m - Y \tag{4.139}$$

We are led to guess the composite function

$$\text{All N:} \quad \frac{dY}{dN} = kY(Y_m - Y) \tag{4.140}$$

which is the logistic differential equation. Equation [4.140] has the properties

$$Y \to 0 \quad \frac{dY}{dN} \to kYY_m = cY \tag{4.141}$$

$$Y \to Y_m \quad \frac{dY}{dN} \to kY_m(Y_m - Y) = c(Y_m - Y) \tag{4.142}$$

where $c = kY_m$. It follows that Eq. [4.140] reduces to the simple exponential form at low Y and to the Mitscherlich form at high Y. Now it can be shown by partial fractions that

$$\frac{Y_m \, dY}{Y(Y_m - Y)} = \frac{dY}{Y} + \frac{dY}{Y_m - Y} = k \, Y_m \, dN = c \, dN \tag{4.143}$$

Integration of Eq. [4.143] leads to

$$\frac{Y}{Y_m - Y} = \left(\frac{Y_0}{Y_m - Y_0}\right) \exp(-cN) \tag{4.144}$$

Equation [4.144] can be rearranged to the regression form

$$Y = \frac{A}{1 + \exp(b - cN)} \tag{4.145}$$

where A, b, c are regression coefficients, and $b = \ln(Y_m/Y_0 - 1)$. The logistic model exhibits sigmoid behavior characteristic of data discussed in Chapters 1 and 2. It is an example of a well-behaved function (Rojansky, 1938, p. 11).

The logistic model possesses a symmetry property. Note that Eq. [4.145] can be written in the dimensionless form

$$\phi = \frac{1}{1 + \exp(-\xi)} \qquad [4.146]$$

where the dimensionless variables are $\phi = Y/A$ and $\xi = cN - b$. It follows that the logistic differential equation is now given by

$$\frac{d\phi}{d\xi} = \phi(1 - \phi) = \frac{1}{\exp(+\xi) + 2 + \exp(-\xi)} \qquad [4.147]$$

which exhibits bilateral symmetry about $\xi = 0$ since the sign on ξ can be changed without altering Eq. [4.147]. Symmetry in physics has proven to be of central importance because it is linked to conservation of various quantities in classical, quantum, and particle physics (Barrow, 1994, p. 160; Bunch, 1989, p. 95; Feynman, 1965; Icke, 1999, p. 106; Jean and Barabé, 1998; Pagels, 1985, p. 188; Rothman, 1985, p. 222; Weinberg, 1993, p. 136; Weyl, 1952). What is conserved in the present system? According to Eq. [4.140] differential change in dry matter with change in applied N is proportional to the filled capacity of the system Y and the unfilled capacity of the system $Y_m - Y$. Now the total capacity of the system is the sum of filled and unfilled components, with total capacity given by

$$\text{Total capacity} = Y + (Y_m - Y) = Y_m = A = \text{constant} \qquad [4.148]$$

This of course assumes that factors which influence A (such as water availability, harvest interval, other nutrients) are all held constant over various levels of applied N. While this result may seem almost trivial, it does illustrate an association of mathematical symmetry and conservation of a quantity.

4.5 COUPLING AMONG APPLIED, SOIL, AND PLANT COMPONENTS

In previous sections we focused on coupling between applied and plant components (above ground vegetation) of the system. In this section we examine data which bring in the soil component, such as extractable elements and plant roots. This will in fact add strength to the modeling approach described above. Overman (1995b) has summarized this work.

Adams and Stelly (1962) and Adams et al. (1966) reported response of coastal bermudagrass to applied P and K, including extractable soil P and K, on Cecil sandy loam. Data are given in Tables 4.1 and 4.2. Response to applied P is shown in Fig. 4.4, where the curves are drawn from

Table 4.1   Plant P uptake and soil extractable P response
to applied P with coastal bermudagrass grown at
Watkinsville, GA (1955–1957)

Applied P kg ha^{-1}	P uptake kg ha^{-1}	Extractable P kg ha^{-1}	Extractable/uptake g g^{-1}
0	16	22	1.4
24	22	46	2.1
48	23	81	3.5
96	26	97	3.7

Data adapted from Adams and Stelly (1962) and Adams et al.
(1966).

$$P_u = \frac{25.0}{1 + \exp(-0.60 - 0.0565P)} \qquad [4.149]$$

$$P_e = \frac{100}{1 + \exp(1.39 - 0.0565P)} \qquad [4.150]$$

$$\frac{P_e}{P_u} = 4.00 \left[\frac{1 + \exp(-0.60 - 0.0565P)}{1 + \exp(1.39 - 0.0565P)} \right] \qquad [4.151]$$

where $P =$ applied   P,   kg ha^{-1}; $P_u =$ plant   P   uptake,   kg ha^{-1};
$P_e =$ extractable soil P, kg ha^{-1}. Overman (1995b) showed that $c = 0.0565$
ha kg^{-1} is statistically the same for Eqs. [4.149] and [4.150]. The overall
correlation coefficient is 0.9983. As a consequence of the common c
parameter, plant P uptake and extractable soil P are related through the
hyperbolic equation

Table 4.2   Plant K uptake and soil extractable K response to
applied K with coastal bermudagrass grown at Watkinsville,
GA (1955–1957).

Applied K kg ha^{-1}	K uptake kg ha^{-1}	Extractable K kg ha^{-1}	Extractable/uptake g g^{-1}
0	106	25	0.22
46	151	33	0.22
92	206	48	0.23
184	291	69	0.24

Data adapted from Adams and Stelly (1962) and Adams et al.
(1966).

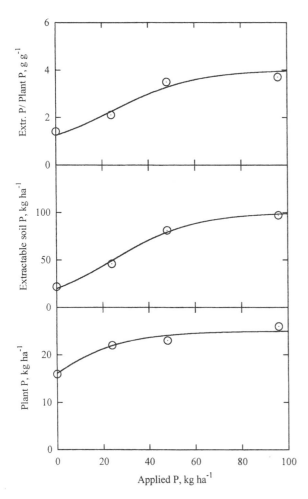

Figure 4.4 Dependence of plant P and extractable soil P on applied P for coastal bermudagrass on Cecil sandy loam. Data from Adams and Stelly (1962) and Adams et al. (1966) as discussed by Overman (1995). Curves drawn from Eqs. [4.149] through [4.151].

$$P_u = \frac{29.0 P_e}{15.8 + P_e}$$

[4.152]

which leads to the linear relationship

$$\frac{P_e}{P_u} = 0.545 + 0.0345 P_e$$

[4.153]

as shown in Fig. 4.5. This confirms the close relationship between extractable soil P and plant P uptake. A similar analysis for applied K shown in Fig. 4.6 leads to

$$K_u = \frac{360}{1 + \exp(0.886 - 0.0127K)} \qquad [4.154]$$

$$K_e = \frac{86.7}{1 + \exp(0.977 - 0.0127K)} \qquad [4.155]$$

$$\frac{K_e}{K_u} = 4.15 \left[\frac{1 + \exp(0.886 - 0.0127K)}{1 + \exp(0.977 - 0.0127K)}\right] \qquad [4.156]$$

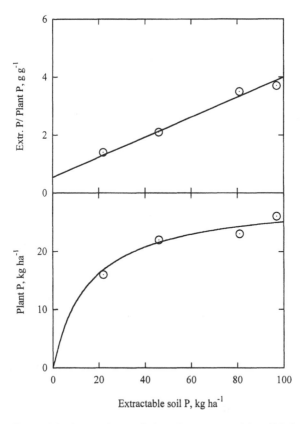

Figure 4.5 Dependence of plant P on extractable soil P for coastal bermudagrass on Cecil sandy loam for the study of Adams and Stelly (1962) and Adams et al. (1966). Curve drawn from Eq. [4.152], line from Eq. [4.153].

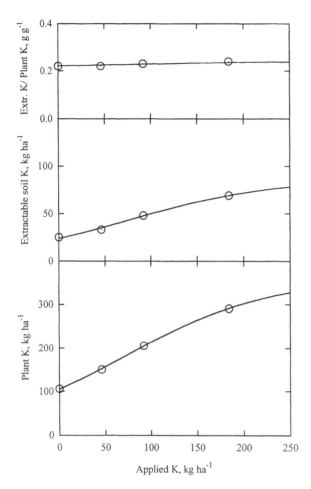

Figure 4.6 Dependence of plant K and extractable soil K on applied K for coastal bermudagrass on Cecil sandy loam. Data from Adams and Stelly (1962) and Adams et al. (1966) as discussed by Overman (1995). Curves drawn from Eqs. [4.154] through [4.156].

where $K =$ applied K, $\mathrm{kg\,ha^{-1}}$; $K_u =$ plant K uptake, $\mathrm{kg\,ha^{-1}}$; $K_e =$ extractable soil K, $\mathrm{kg\,ha^{-1}}$. The overall correlation coefficient is 0.9999. Again, as a consequence of the common c parameter, plant K uptake and extractable soil K are related through the hyperbolic equation

$$K_u = \frac{4140K_e}{910 + K_e} \qquad\qquad [4.157]$$

which leads to the linear relationship

$$\frac{K_e}{K_u} = 0.220 + 0.000242K_e \qquad\qquad [4.158]$$

as shown in Fig. 4.7. This confirms the close relationship between extractable soil K and plant K uptake. Equation [4.157] is indicative of luxury uptake of K by the grass.

Wilkinson and Mays (1979) studied response of plant roots and tops to applied N for KY 31 tall fescue (*F. arundinacea* Schreb.) at Watkinsville, GA and Saluda, SC. Data are listed in Table 4.3 and shown in Fig. 4.8, where the curves are drawn from

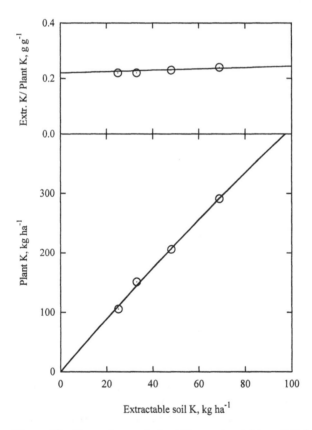

Figure 4.7 Dependence of plant K on extractable soil K for coastal bermudagrass on Cecil sandy loam for the study of Adams and Stelly (1962) and Adams et al. (1966). Curve drawn from Eq. [4.157], line from Eq. [4.158].

Table 4.3 Dry matter responses to applied N for plant tops and roots for KY 31 tall fescue grown at Watkinsville, GA and Saluda, SC

Applied N kg ha^{-1}	Plant dry matter, Mg ha^{-1}		Tops/roots g g^{-1}
	Tops	Roots	
0	1.2	2.57	0.47
112	4.6	4.04	1.14
224	7.4	4.64	1.59
448	10.3	5.10	2.02
896	12.2	—	—

Data adapted from Wilkinson and Mays (1979).

$$Y_r = \frac{5.22}{1 + \exp(0.00 - 0.0097N)} \qquad [4.159]$$

$$Y_t = \frac{11.7}{1 + \exp(1.67 - 0.0097N)} \qquad [4.160]$$

$$\frac{Y_t}{Y_r} = 2.24 \left[\frac{1 + \exp(0.00 - 0.0097N)}{1 + \exp(1.67 - 0.0097N)} \right] \qquad [4.161]$$

where $N =$ applied N, kg ha^{-1}; $Y_t =$ dry matter yield of plant tops, Mg ha^{-1}; $Y_r =$ dry matter yield of plant roots, Mg ha^{-1}. The overall correlation coefficient is 0.9935. Due to the common $c = 0.0097$ ha kg^{-1}, dry matter yield for roots and tops are related by

$$Y_r = \frac{6.43 Y_t}{2.72 + Y_t} \qquad [4.162]$$

$$\frac{Y_t}{Y_r} = 0.42 + 0.156 Y_t \qquad [4.163]$$

as shown in Fig. 4.9. Equation [4.162] suggests that plant root growth depends upon transfer of some constituent from plant tops. In fact it is known that carbohydrates are manufactured in plant tops by photosynthesis and transferred downward to plant roots.

Further support for the hyperbolic relationship between plant roots and tops (above ground vegetation) is provided in Table 4.4 from the work of Blue (1973) with pensacola bahiagrass. Response of plant tops to applied N is shown in Fig. 4.10, where the curves are drawn from

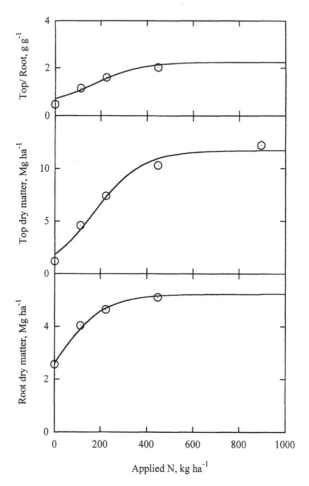

Figure 4.8 Dependence of plant root and plant top dry matter on applied N for tall fescue. Data from Wilkinson and Mays (1979). Curves drawn from Eqs. [4.159] through [4.161].

$$Y_t = \frac{14.2}{1 + \exp(1.34 - 0.0133N)} \tag{4.164}$$

$$N_{ut} = \frac{235}{1 + \exp(1.86 - 0.0133N)} \tag{4.165}$$

$$N_{ct} = 16.5 \left[\frac{1 + \exp(1.34 - 0.0133N)}{1 + \exp(1.86 - 0.0133N)} \right] \tag{4.166}$$

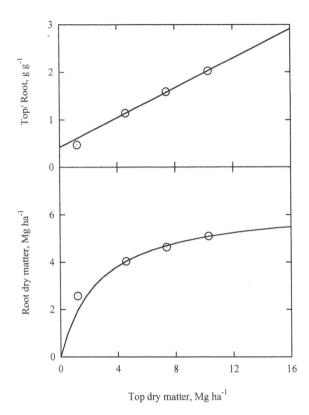

Figure 4.9 Dependence of root dry matter on top dry matter for tall fescue from study of Wilkinson and Mays (1979). Curve drawn from Eq. [4.162]; line drawn from Eq. [4.163].

where N = applied N, $kg\,ha^{-1}$; Y_t = dry matter yield for tops, $Mg\,ha^{-1}$; N_{ut} = plant N uptake for tops, $kg\,ha^{-1}$; N_{ct} = plant N concentration for tops, $g\,kg^{-1}$. The overall correlation coefficient is 0.9971. In this case the hyperbolic equation is

$$Y_t = \frac{35.4 N_{ut}}{350 + N_{ut}} \qquad\qquad [4.167]$$

which leads to the linear relationship

$$N_{ct} = 9.89 + 0.0282 N_{ut} \qquad\qquad [4.168]$$

These results are shown in Fig. 4.11. The corresponding equations for plant roots are

Table 4.4 Dry matter and plant N uptake response to applied N for plant tops and roots for pensacola bahiagrass grown at Gainesville, FL

Component	Applied N kg ha^{-1}	Dry matter Mg ha^{-1}	N uptake kg ha^{-1}	N conc. g kg^{-1}
Tops	0	3.2	37	11.6
	112	7.3	90	12.3
	224	12.4	178	14.4
	448	13.6	232	17.1
Roots	0	8.6	47	5.5
	112	15.6	89	5.7
	224	17.2	138	8.0
	448	13.4	218	16.3

Data adapted from Blue (1973).

$$Y_r = \frac{18.5}{1 + \exp(0.00 - 0.0133N)} \qquad [4.169]$$

$$N_{ur} = \frac{175}{1 + \exp(1.34 - 0.0133N)} \qquad [4.170]$$

$$N_{cr} = 9.46 \left[\frac{1 + \exp(0.00 - 0.0133N)}{1 + \exp(1.34 - 0.0133N)} \right] \qquad [4.171]$$

where N = applied N, kg ha^{-1}; Y_t = dry matter yield for roots, Mg ha^{-1}; N_{ur} = plant N uptake for roots, kg ha^{-1}; N_{cr} = plant N concentration for roots, g kg^{-1}. Results are shown in Fig. 4.12. Values for $N = 448$ kg ha^{-1} have been ignored in the analysis. The hyperbolic and linear equations are

$$Y_r = \frac{25.1 N_{ur}}{62.1 + N_{ur}} \qquad [4.172]$$

$$N_{cr} = 2.47 + 0.0398 N_{ur} \qquad [4.173]$$

as shown in Fig. 4.13. As a consequence of Eqs. [4.164] and [4.169], dry matter for plant roots and tops can be related through

$$Y_r = \frac{25.1 Y_t}{5.04 + Y_t} \qquad [4.174]$$

$$\frac{Y_t}{Y_r} = 0.20 + 0.0398 Y_t \qquad [4.175]$$

as shown in Fig. 4.14. Similarly, combination of Eqs. [4.165] and [4.170] leads to

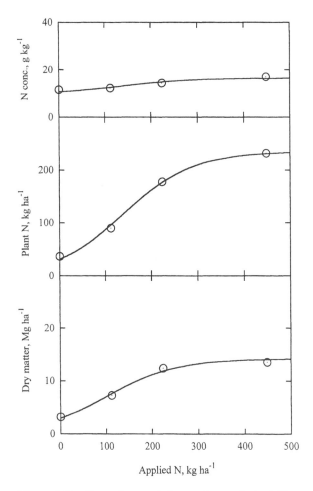

Figure 4.10 Dependence of dry matter and plant N on applied N for pensacola bahiagrass tops. Data from Blue (1973). Curves drawn from Eqs. [4.164] through [4.166].

$$N_{ur} = \frac{432N_{ut}}{345 + N_{ut}} \tag{4.176}$$

$$\frac{N_{ut}}{N_{ur}} = 0.80 + 0.00232N_{ut} \tag{4.177}$$

as shown in Fig. 4.15.

This analysis has confirmed coupling among various components of the plant/soil system. The subject of nutrient uptake from soil solution has been

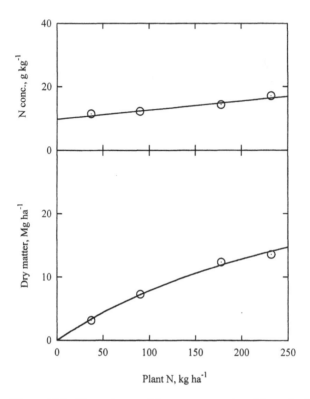

Figure 4.11 Dependence of dry matter on plant N uptake for pensacola bahiagrass tops from study of Blue (1973). Curve drawn from Eq. [4.167]; line drawn from Eq. [4.168].

discussed by Barber (1984) and by Mengel and Kirkby (1987, p. 144). Linkage of plant carbon between plant tops and roots through photosynthesis is well known (Marschner, 1986, p. 446). Flow of organic N from plant tops to roots is related to the flow of inorganic N from soil solution to plant roots then to plant tops.

Two hyperbolic relationships have emerged. The first links plant tops T to available soil nutrients S by

$$T = \frac{aS}{b + S} \qquad [4.178]$$

where a, b are constants for a given system. The second links plant roots R to plant tops by

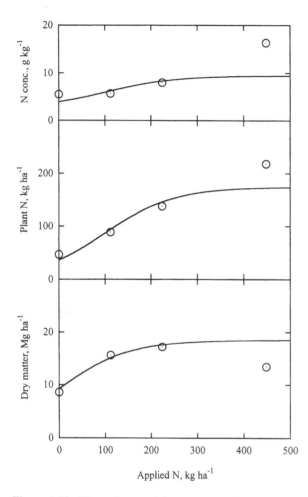

Figure 4.12 Dependence of dry matter and plant N on applied N for pensacola bahiagrass roots. Data from Blue (1973). Curves drawn from Eqs. [4.169] through [4.171].

$$R = \frac{cT}{d + T}$$ [4.179]

where c, d are constants. Substitution of Eq. [4.178] into Eq. [4.179] leads to

$$R = \frac{fS}{g + S}$$ [4.180]

where the constants are related through

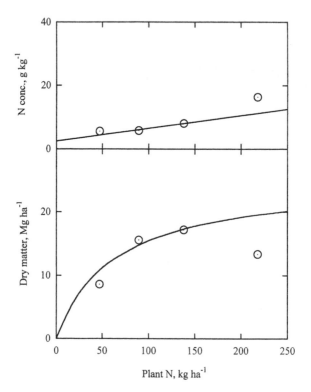

Figure 4.13 Dependence of dry matter on plant N uptake for pensacola bahiagrass roots from study of Blue (1973). Curve drawn from Eq. [4.172]; line drawn from Eq. [4.173].

$$f = \frac{ac}{a+d} \qquad\qquad [4.181]$$

$$g = \frac{bd}{a+d} \qquad\qquad [4.182]$$

While Eq. [4.180] has not been confirmed directly with data, it suggests that root constituents (carbon or nitrogen compounds) are related to available soil nutrients, which in turn are related to applied nutrients.

We now formalize this procedure as a theorem. The theorem is generalized to include both hyperbolic and logistic functions.

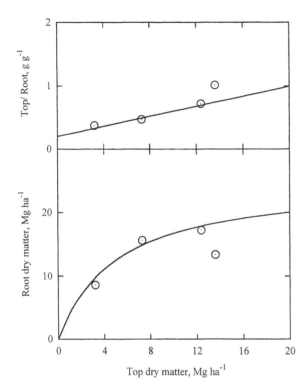

Figure 4.14 Dependence of root dry matter on top dry matter for pensacola bahia-grass from study of Blue (1973). Curve drawn from Eq. [4.174]; line drawn from Eq. [4.175].

Theorem. The hyperbolic transform of a hyperbolic function leads to another hyperbolic function. The hyperbolic transform of a logistic function leads to another logistic function.

Proof. Define the hyperbolic transform for the variable $y \geq 0$ by

$$H\{y\} = \frac{\alpha y}{\beta + y} \quad \alpha, \beta = \text{constants} \geq 0 \qquad [4.183]$$

where the transform of variable y is designated by the symbol $H\{y\}$.

Case 1. Consider a hyperbolic function for the independent variable $x \geq 0$

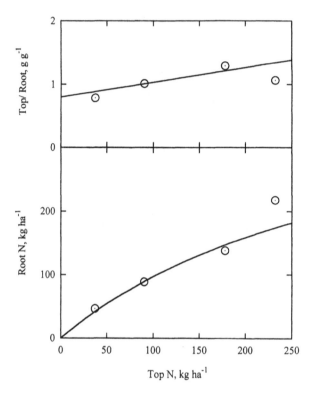

Figure 4.15 Dependence of root N on top N for pensacola bahiagrass from study of Blue (1973). Curve drawn from Eq. [4.176]; line drawn from Eq. [4.177]

$$y = \frac{Ax}{K' + x} \qquad A, K' = \text{constants} \geq 0 \qquad\qquad [4.184]$$

The hyperbolic transform of Eq. [4.184] leads to

$$H\{y\} = \frac{\alpha y}{\beta + y} = \frac{\alpha \dfrac{Ax}{K' + x}}{\beta + \dfrac{Ax}{K' + x}} = \frac{\alpha A x}{\beta K' + (\beta + A)x} = \frac{\dfrac{\alpha A}{\beta + A}x}{\dfrac{\beta K'}{\beta + A} + x} = \frac{\alpha' x}{\beta' + x} \geq 0$$

$$[4.185]$$

which is also a hyperbolic equation with the new constants

$$\alpha' = \frac{A\alpha}{\beta + A} \qquad \text{and} \qquad \beta' = \frac{K'\beta}{\beta + A} \qquad\qquad [4.186]$$

Since the parameters α, β, A, and K' are all assumed positive, it follows that the new parameters α' and β' are also positive.

Case 2.   Consider a logistic function for the independent variable $-\infty < x < +\infty$ defined by

$$y = \frac{A}{1 + \exp(b - cx)} \geq 0 \qquad A, c = \text{constants} \geq 0, b = \text{constant} \quad [4.187]$$

The hyperbolic transform of Eq. [4.187] leads to

$$H\{y\} = \frac{\alpha y}{\beta + y} = \frac{\alpha \dfrac{A}{1 + \exp(b - cx)}}{\beta + \dfrac{A}{1 + \exp(b - cx)}} = \frac{\alpha A}{(\beta + A) + \beta \exp(b - cx)}$$

$$= \frac{\dfrac{\alpha A}{\beta + A}}{1 + \left(\dfrac{\beta}{\beta + A}\right)\exp(b - cx)} = \frac{A'}{1 + \exp(b' - cx)} \geq 0 \qquad [4.188]$$

which is also a logistic equation with the new constants given by

$$A' = \frac{\alpha A}{\beta + A} \qquad \text{and} \qquad \exp(b') = \frac{\beta}{\beta + A}\exp(b) \qquad [4.189]$$

The second relation in Eq. [4.189] can be rewritten as

$$b' = b + \ln\left(\frac{\beta}{\beta + A}\right) = b - \ln\left(1 + \frac{A}{\beta}\right) \qquad [4.190]$$

Since we have assumed A and β positive, it follows that $A/\beta > 0$, so that $b' < b$. Both b and b' are dimensionless parameters. Parameter c is unchanged by the transformation. It has the reciprocal units of x.

4.6 ACCUMULATION OF DRY MATTER AND PLANT NUTRIENTS

In Chapter 3 a hyperbolic relationship was established between dry matter and plant N accumulation with time (Eq. [3.63]), viz.

$$N_u = \frac{N_{um} Y}{K_y + Y} \qquad [4.191]$$

where N_u = plant N accumulation, kg ha^{-1}; Y = dry matter accumulation, Mg ha^{-1}; N_{um} = maximum plant N accumulation, kg ha^{-1}; K_y = yield response coefficient, Mg ha^{-1}. We now develop an approximate derivation of Eq. [4.191]. First write down an equation for net rate of accumulation, including gain and loss:

$$\frac{dN_u}{dt} = k_f Y (N_{um} - N_u) - k_b N_u \qquad [4.192]$$

where k_f and k_b are forward (gain) and backward (loss) rate coefficients, respectively. Then use the quasiequilibrium assumption (similar to enzyme kinetics), $dN_u/dt \approx 0$, to obtain

$$k_f N_{um} Y - k_f Y N_u - k_b N_u \approx 0 \qquad [4.193]$$

or

$$N_u = \frac{N_{um} Y}{k_b/k_f + Y} = \frac{N_{um} Y}{K_y + Y} \qquad [4.194]$$

where $K_y = k_b/k_f$ = dry matter response coefficient, Mg ha^{-1}. This model predicts that if accumulation shuts off ($k_f = 0$), then N_u will decay exponentially according to the equation

$$N_u = N_{ui} \exp[-k_b(t - t_i)] \qquad [4.195]$$

where t_i = time of initiation of decay, wk; N_{ui} = plant nutrient accumulation at time of initiation of decay, kg ha^{-1}. Equation [4.192] treats plant N accumulation with time as a dynamic process similar to autocatalysis in chemistry (Johnson et al., 1974; Moore and Pearson, 1981), enzyme kinetics (Laidler, 1965), and phosphorus kinetics in soil (Overman and Scholtz, 1999).

Evidence for just such an occurrence as described by Eq. [4.195] was presented in Section 1.3 for potassium accumulation by wheat at Essex, England. For that case Eq. [4.195] becomes

$$K_u = K_{ui} \exp[-k_b(t - t_i)] = 160 \exp[-0.094(t - 23.5)] \qquad [4.196]$$

From Eqs. [1.21] and [1.23], maximum $Q_m = 3.46$ and maximum $Y_m = 17.75$ Mg ha^{-1}. It follows from Eq. [1.31] that $K_y = 10.7$ Mg ha^{-1} and that maximum K uptake would be $K_{um} = 235$ kg ha^{-1}. Since from Eq. [4.196] $k_b = 0.094$ wk^{-1} and $K_y = k_b/k_f = 10.7$ Mg ha^{-1}, we obtain $k_f = 0.0088$ ha Mg^{-1} wk^{-1}. At initiation of decay plant K was $K_{ui} = 160$ kg ha^{-1}, so plant K accumulation reached $160/235 = 68\%$ of projected maximum. Apparently something happened in the plant system to shut off the accumulation process.

EXERCISES

4.1 Sartain (1992) measured dependence of extractable soil P on applied P for Arredondo fine sand (loamy, siliceous hyperthermic Grossarenic Paleudult) at Gainesville, FL. Data are given in the table below.

a. Plot extractable soil phosphorus P_e vs. applied P on linear graph paper.

b. Plot $A/P_e - 1$ vs. P on semilog graph paper.

c. Perform linear regression on $\ln(A/P_e - 1)$ vs. P for the values from the table.

d. Nonlinear regression of the logistic equation for these data leads to

$$P_e = \frac{502}{1 + \exp(0.85 - 0.0330P)}$$

Calculate and draw the response curve on (a) from this equation.

e. Discuss the results.

Dependence of extractable soil P on applied P for study of Sartain (1992)

Applied P $g\,m^{-2}$	Extractable P $mg\,kg^{-1}$	$A/P_e - 1$ $A = 500$	$A = 550$	$A = 502$
0	151	2.31	2.64	2.32
10	192	1.60	1.86	1.61
20	218	1.29	1.52	1.30
40	313	0.597	0.757	0.604
80	429	0.165	0.282	0.170

Data adapted from Sartain (1992).

4.2 Vicente-Chandler et al. (1959) measured response of tropical grasses to applied N and harvest interval at Rio Piedras, Puerto Rico. Results are given below for guineagrass (*Panicum maximum*) grown on Fajardo clay soil.

a. Plot dry matter Y, plant N uptake N_u, and plant N concentration N_c vs. applied N on linear graph paper for the three harvest intervals Δt.

b. Draw the response curves on (a) from the logistic equations

$$Y = \frac{A}{1 + \exp(0.50 - 0.0030N)}$$

$$N_u = \frac{A_n}{1 + \exp(1.50 - 0.0030N)}$$

$$N_c = \frac{A_n}{A}\left[\frac{1 + \exp(0.50 - 0.0030N)}{1 + \exp(1.50 - 0.0030N)}\right]$$

where $A(40) = 32$, $A(60) = 40$, $A(90) = 54\,\mathrm{Mg\,ha^{-1}}$; $A_n(40) = 720$, $A_n(60) = 740$, $A_n(90) = 800\,\mathrm{kg\,ha^{-1}}$.

   c. Show the phase plot for Y vs. N_u and N_c vs. N_u for the three harvest intervals on linear graph paper.
   d. Calculate and plot the phase relations on (c).
   e. Perform linear regression on A vs. Δt and A_n vs. Δt.
   f. Plot results for (e) on linear graph paper.
   g. Discuss the results.

Response of dry matter and plant N uptake to applied N for guineagrass for study of Vicente-Chandler et al. (1959)

Harvest interval days	Applied N kg ha^{-1}	Dry Matter Mg ha^{-1}	Plant N uptake kg ha^{-1}	Plant N conc. g kg^{-1}
40	0	10.3	120	12.0
	224	16.8	210	13.0
	448	26.7	380	14.0
	896	30.9	540	17.8
	1792	30.0	695	23.5
60	0	12.2	120	10.0
	224	23.9	230	10.0
	448	32.7	370	11.0
	896	36.7	550	15.0
	1792	38.5	760	19.8
90	0	14.8	120	7.8
	224	34.7	280	8.0
	448	41.3	370	9.0
	896	48.8	580	12.0
	1792	54.1	860	16.0

Data adapted from Vicente-Chandler et al. (1959).

4.3 Vicente-Chandler et al. (1959) measured response of tropical grasses to applied N and harvest interval at Rio Piedras, Puerto Rico. Results are given below for paragrass (*Panicum purpurascens*) grown on Fajardo clay soil.

a. Plot dry matter Y, plant N uptake N_u, and plant N concentration N_c vs. applied N on linear graph paper for the three harvest intervals Δt.

b. Draw the response curves on (a) from the logistic equations

$$Y = \frac{A}{1 + \exp(0.50 - 0.0030N)}$$

$$N_u = \frac{A_n}{1 + \exp(1.50 - 0.0030N)}$$

$$N_c = \frac{A_n}{A} \left[\frac{1 + \exp(0.50 - 0.0030N)}{1 + \exp(1.50 - 0.0030N)} \right]$$

where $A(40) = 28$, $A(60) = 37$, $A(90) = 45 \, \text{Mg ha}^{-1}$; $A_n(40) = 610$, $A_n(60) = 630$, $A_n(90) = 670 \, \text{kg ha}^{-1}$.

c. Show the phase plot for Y vs. N_u and N_c vs. N_u for the three harvest intervals on linear graph paper.

d. Calculate and plot the phase relations on (c).

e. Perform linear regression on A vs. Δt and A_n vs. Δt.

f. Plot results for (e) on linear graph paper.

g. Discuss the results.

Response of dry matter and plant N uptake to applied N for paragrass for study of Vicente-Chandler et al. (1959)

Harvest interval days	Applied N kg ha^{-1}	Dry Matter Mg ha^{-1}	Plant N uptake kg ha^{-1}	Plant N conc. g kg^{-1}
40	0	5.8	70	12.0
	224	13.0	171	13.0
	448	19.2	284	15.0
	896	26.4	494	18.7
	1792	28.5	684	24.0
60	0	8.7	87	10.0
	224	19.7	197	10.0
	448	26.7	306	12.0
	896	34.5	515	15.0
	1792	36.5	708	19.4
90	0	14.6	110	7.7
	224	25.5	180	6.9
	448	36.9	280	7.7
	896	40.6	490	12.0
	1792	43.9	610	14.0

Data adapted from Vicente-Chandler et al. (1959).

4.4 McEwen et al. (1989) reported data for ryegrass (*Lolium perenne*) response to applied N. The soil at Rothamsted was silty clay loam, while the soil at Woburn was sandy loam. Data are given below for the two sites.
  a. Plot the data for Rothamsted on linear graph paper.
  b. Plot model curves on (a) from the equations

$$Y = \frac{14.1}{1 + \exp(1.80 - 0.0100N)}$$

$$N_u = \frac{410}{1 + \exp(2.50 - 0.0100N)}$$

$$N_c = 29.1 \left[\frac{1 + \exp(1.80 - 0.0100N)}{1 + \exp(2.50 - 0.0100N)} \right]$$

  c. Plot Y and N_c vs. N_u on linear graph paper.
  d. Calculate and plot the phase relations on (c).
  e. Repeat (a) through (d) with the same logistic equations for the Woburn site.
  f. Compare results for the two sites.

Dependence of dry matter and plant N on applied N for ryegrass (irrigated, 6 cuttings) at Rothamsted Experiment Station, England

N kg ha^{-1}	Y Mg ha^{-1}	N_u kg ha^{-1}	N_c g kg^{-1}
0	2.0	40	18
100	3.6	70	18
200	7.8	160	20
300	10.0	210	21
400	13.4	340	26
500	13.7	390	29
600	13.9	450	32

Data adapted from McEwen et al. (1989)

Dependence of dry matter and plant N
on applied N for ryegrass (irrigated, 6
cuttings) at Woburn Experiment
Station, England

N kg ha^{-1}	Y Mg ha^{-1}	N_u kg ha^{-1}	N_c g kg^{-1}
0	1.4	20	18
100	3.8	70	18
200	6.5	130	20
300	10.8	230	21
400	13.0	310	24
500	14.7	400	28
600	14.3	420	30

Data adapted from McEwen et al. (1989)

4.5 Baker et al. (1990) measured response of rice (*Oryza sativa* L.) to CO_2 in an outdoor controlled environment chamber with Zolfo fine sand (sandy, siliceous, hyperthermic, Grossarenic Entic Haplohumods) soil. Data are given in the table below.
a. Plot dry matter vs. CO_2 on linear graph paper.
b. Plot $3.5/Y - 1$ vs. CO_2 on semilog paper.
c. Perform linear regression on $\ln 3.5/Y - 1$ vs. CO_2. Plot the best fit line on (b).
d. Draw the model curve on (a) from the logistic equation

$$Y = \frac{3.50}{1 + \exp(1.50 - 0.0050\,CO_2)}$$

e. Estimate the level of CO_2 to produce 50% of maximum yield.

Grain yield of rice in response to
carbon dioxide concentrations at
Gainesville, FL

CO_2 μmol mol^{-1}	Dry matter g	$3.5/Y - 1$
160	1.4	1.5
250	1.3	1.7
330	1.9	0.84
500	3.0	0.17
660	2.8	0.25
900	3.3	0.061

Data adapted from Baker et al. (1990).

4.6 Baker et al. (1990) also measured accumulation of dry matter with time of rice by roots and tops (above ground) in response to CO_2. Data are given in the table below.

   a. Plot dry matter for roots Y_r, tops Y_t, and the ratio of tops/roots Y_t/Y_r vs. CO_2 on linear graph paper for 110 days after planting.

   b. Draw the curves on (a) from the model equations

$$Y_r = \frac{4.0}{1 + \exp(1.00 - 0.0050\, CO_2)}$$

$$Y_t = \frac{100}{1 + \exp(1.50 - 0.0050\, CO_2)}$$

$$\frac{Y_t}{Y_r} = 25.0 \left[\frac{1 + \exp(1.00 - 0.0050\, CO_2)}{1 + \exp(1.50 - 0.0050\, CO_2)} \right]$$

   c. Calculate the phase parameters Y_{rm} and K'.

   d. Plot Y_r and Y_t/Y_r vs. Y_t on linear graph paper. Draw the corresponding curve and line as well.

   e. Discuss the results, particularly the scatter of data points.

Response of rice to carbon dioxide concentrations at Gainesville, FL

CO_2	Top dry matter g			Root dry matter g			Top/Root dry matter g		
μmol mol^{-1}	\multicolumn{9}{c}{Days after planting}
	30	58	110	30	58	110	30	58	110
160	6.2	15	35	0.4	0.6	1.7	15	24	21
250	8.5	31	43	0.5	1.8	2.6	17	17	16
330	9.1	33	65	1.0	1.6	2.2	9.1	21	29
500	12	41	89	1.3	3.9	3.2	9.0	10	28
660	14	47	84	1.9	2.9	4.3	7.6	16	20
900	14	51	76	1.9	4.0	3.2	7.3	13	24

Data adapted from Baker et al. (1990).

4.7 Alvarez-Sánchez et al. (1999) conducted a study on response of potato (*Solanum tuberosum*) to applied phosphorus on a clay loam soil. Data for dry matter, plant P, and extractable soil P are given in the table below.

   a. Plot dry matter Y, plant P uptake P_u, and plant P concentration P_c vs. applied phosphorus P on linear graph paper.

   b. Draw simulation curves on (a) from the model

$$Y = \frac{5.7}{1 + \exp(-0.7 - 0.0125\,P)}$$

$$P_u = \frac{12}{1 + \exp(-0.5 - 0.0125\,P)}$$

$$P_c = 2.1 \left[\frac{1 + \exp(-0.7 - 0.0125\,P)}{1 + \exp(-0.5 - 0.0125\,P)}\right]$$

c. Plot P_u , extractable soil P, P_e, and P_e/P_u vs. P on linear graph paper.

d. Draw simulation curves on (c) from the model

$$P_u = \frac{12}{1 + \exp(-0.5 - 0.0125\,P)}$$

$$P_e = \frac{23}{1 + \exp(+0.9 - 0.0125\,P)}$$

$$\frac{P_e}{P_u} = 1.9 \left[\frac{1 + \exp(-0.5 - 0.0125\,P)}{1 + \exp(+0.9 - 0.0125\,P)}\right]$$

e. Plot P_u and P_e/P_u vs. P_e on linear graph paper.

f. Draw the curve and line on (e) from the model

$$P_u = \frac{16.6\,P_e}{7.9 + P_e}$$

$$\frac{P_e}{P_u} = 0.474 + 0.0601\,P_e$$

g. Discuss connections between the components of the plant/soil system.

Response of potato dry matter, plant P uptake, and soil extractable P to applied P at Valle de Mexico

P kg ha^{-1}	Y Mg ha^{-1}	P_u kg ha^{-1}	P_c g kg^{-1}	P_e µg g^{-1}
0	2.7	5.6	2.1	6.14
18	4.4	9.7	2.2	9.25
41	4.2	8.8	2.1	7.60
46	4.4	8.7	2.0	9.18
69	4.7	9.4	2.0	9.25
78	5.0	11.0	2.2	11.9
90	5.1	10.2	2.0	13.4
106	5.4	10.8	2.0	13.0

continues

continued

P kg ha^{-1}	Y Mg ha^{-1}	P_u kg ha^{-1}	P_c g kg^{-1}	P_e μg g^{-1}
113	5.0	10.5	2.1	11.8
135	5.2	11.4	2.2	14.7
150	5.0	10.5	2.1	15.8
163	5.4	11.4	2.1	18.3
207	5.4	11.9	2.2	22.1

Data adapted from Alvarez-Sánchez et al. (1999).

4.8 Kamuru et al. (1998) reported response of rice (*Oryza sativa* L.) to applied nitrogen with Millhopper fine sand (fine-loamy, siliceous, hyperthermic Grossarenic Paleudults) in a growth chamber at Belle Glade, FL. Data are given in the table below for tops (shoots + grain) and roots for uninoculated plants for 1990.
   a. Plot dry matter Y, plant N uptake N_u, and plant N concentration N_c vs. applied nitrogen N on linear graph paper.
   b. Draw the model curves on (a) from

$$Y = \frac{A}{1 + \exp(0.00 - 0.0300N)}$$

$$N_u = \frac{A_n}{1 + \exp(0.50 - 0.0300\ N)}$$

$$N_c = N_{cm}\left[\frac{1 + \exp(0.00 - 0.0300N)}{1 + \exp(0.50 - 0.0300N)}\right]$$

where A (tops) = 100 g pot^{-1}, A (roots) = 32 g pot^{-1}; A_n (tops) = 900 mg pot^{-1}; A_n (roots) = 300 mg pot^{-1}; N_{cm} (tops) = 9.0 mg g^{-1}; N_{cm} (roots) = 9.4 mg g^{-1}.
   c. Plot Y and N_c vs. N_u on linear graph paper.
   d. Plot the curve and line on (c) from the equations

$$Y = \frac{254N_u}{1390 + N_u} \quad \text{(tops)}$$

$$N_c = 5.47 + 0.00394N_u \quad \text{(tops)}$$

$$Y = \frac{81.3N_u}{462 + N_u} \quad \text{(roots)}$$

$$N_c = 5.68 + 0.0123N_u \quad \text{(roots)}$$

   e. Plot Y_r and Y_t/Y_r vs. Y_t on linear graph paper.
   f. Draw the lines on (e) from the equations

$$Y_r = 0.339 \, Y_t$$

$$\frac{Y_t}{Y_r} = 2.95$$

g. Plot N_{ur} and N_{ut}/N_{ur} vs. N_{ut} on linear graph paper.

h. Draw the lines on (g) from the equations

$$N_{ur} = 0.363 \, N_{ut}$$

$$\frac{N_{ut}}{N_{ur}} = 2.75$$

i. Discuss the results. How well do the logistic equations fit the data? Why are the phase plots between roots and tops for dry matter and plant N uptake linear?

Response of tops and roots to applied nitrogen for rice grown at Belle Glade, FL.

Component	N kg ha^{-1}	Y g pot^{-1}	N_u mg pot^{-1}	N_c mg g^{-1}
Tops	0	51.2	340	6.64
	25	65.0	500	7.69
	50	81.9	647	7.90
	100	96.1	832	8.66
Roots	0	18.3	137	7.49
	25	21.1	166	7.87
	50	27.9	239	8.57
	100	32.5	303	9.32

Data adapted from Kamuru et al. (1998).

4.9 Beaty et al. (1980) reported response of bahiagrass to applied nitrogen with Eustis loamy sand (sandy, siliceous, thermic Psammentic Paleudults) at Americus, GA. Data are given in the table below for tops (above ground vegetation) and roots.

a. Plot dry matter Y, plant N uptake N_u, and plant N concentration N_c vs. applied nitrogen N for tops on linear graph paper.

b. Draw the model curves on (a) from

$$Y_t = \frac{10.78}{1 + \exp(1.15 - 0.0210N)}$$

$$N_{ut} = \frac{212}{1 + \exp(1.75 - 0.0210\,N)}$$

$$N_{ct} = 19.7 \left[\frac{1 + \exp(1.15 - 0.0210N)}{1 + \exp(1.75 - 0.0210N)} \right]$$

c. Plot Y_t and N_{ct} vs. N_{ut} for tops on linear graph paper.
d. Plot the curve and line on (c) from the equations

$$Y_t = \frac{23.9 N_{ut}}{258 + N_{ut}} \quad \text{(tops)}$$

$$N_{ct} = 10.8 + 0.0419 N_{ut} \quad \text{(tops)}$$

e. Plot dry matter Y, plant N uptake N_u, and plant N concentration N_c vs. applied nitrogen N for roots on linear graph paper.
f. Draw the model curves on (e) from

$$Y_r = \frac{11.50}{1 + \exp(0.75 - 0.0210N)}$$

$$N_{ur} = \frac{192}{1 + \exp(1.35 - 0.0210\,N)}$$

$$N_{cr} = 16.7 \left[\frac{1 + \exp(0.75 - 0.0210N)}{1 + \exp(1.35 - 0.0210N)} \right]$$

g. Plot Y_r and N_{cr} vs. N_{ur} for roots on linear graph paper.
h. Plot the curve and line on (g) from the equations

$$Y_r = \frac{34.9 N_{ur}}{390 + N_{ur}} \quad \text{(roots)}$$

$$N_{cr} = 11.2 + 0.0286 N_{ur} \quad \text{(roots)}$$

i. Plot Y_r and Y_t/Y_r vs. Y_t on linear graph paper.
j. Draw the curve and line on (i) from the equations

$$Y_r = \frac{34.9 Y_t}{21.9 + Y_t}$$

$$\frac{Y_t}{Y_r} = 0.628 + 0.0286 Y_t$$

k. Plot N_{ur} and N_{ut}/N_{ur} vs. N_{ut} on linear graph paper.
l. Draw the lines on (k) from the equations

$$N_{ur} = \frac{580 N_{ut}}{430 + N_{ut}}$$

$$\frac{N_{ut}}{N_{ur}} = 0.741 + 0.00172 N_{ut}$$

m. Discuss the results. How well do the logistic equations fit the data?

Response of tops and roots to applied nitrogen for bahiagrass grown at Americus, GA.

Component	N kg ha^{-1}	Y Mg ha^{-1}	N_u kg ha^{-1}	N_c g kg^{-1}
Tops	0	2.55	36	14.2
	84	6.90	104	15.1
	168	10.20	176	17.3
	336	10.75	211	19.6
Roots	0	3.74	36	9.5
	84	10.52	112	10.6
	168	13.59	173	12.7
	336	11.06	191	17.3

Data adapted from Beaty et al. (1980).

4.10 Burton et al. (1997) reported data for Pensacola bahiagrass (*Paspalum notatum*) response to applied N, P, and K at Tifton, GA. The soil at was Clarendon loamy sand (fine-loamy, siliceous, thermic Plinthaquic Paleudults). Data are given below for 1993, when average harvest interval was 7.1 wk.

a. Plot the data for dry matter Y, plant N uptake N_u, and plant N concentration N_c vs. applied nitrogen N on linear graph paper.

b. Plot model curves on (a) from the equations

$$Y = \frac{16.7}{1 + \exp(1.55 - 0.0125N)}$$

$$N_u = \frac{283}{1 + \exp(2.15 - 0.0125N)}$$

$$N_c = 16.9 \left[\frac{1 + \exp(1.55 - 0.0125N)}{1 + \exp(2.15 - 0.0125N)} \right]$$

c. Plot Y and N_c vs. N_u on linear graph paper.

d. Calculate and plot the phase relations on (c).

e. Discuss the results.

Dependence of dry matter and plant N
on applied N for bahiagrass at Tifton,
GA

N kg ha^{-1}	Y Mg ha^{-1}	N_u kg ha^{-1}	N_c g kg^{-1}
56	4.95	52.5	10.6
112	8.40	95	11.3
224	12.63	173	13.7
448	16.44	273	16.6

Data adapted from Burton et al. (1997).

4.11 Rose and Biernacka (1999) reported data for Freeman maple (*Acer* ×
freemanii E. Murr.) tree response to applied N grown in containers at
Columbus, OH. Data are given below for 1996–1997, for a growth
period of one year.
a. Plot the data for dry matter Y, plant N uptake N_u, and plant N
concentration N_c vs. applied nitrogen N on linear graph paper.
b. Plot model curves on (a) from the equations

$$Y = \frac{65.8}{1 + \exp(1.30 - 1.25N)}$$

$$N_u = \frac{840}{1 + \exp(1.85 - 1.25N)}$$

$$N_c = 12.8 \left[\frac{1 + \exp(1.30 - 1.25N)}{1 + \exp(1.85 - 1.25N)} \right]$$

c. Plot Y and N_c vs. N_u on linear graph paper.
d. Calculate and plot the phase relations on (c).
e. Discuss the results.

Dependence of dry matter and plant N on
applied N for Freeman maple trees at
Columbus, OH

N g plant^{-1}	Y g plant^{-1}	N_u mg plant^{-1}	N_c g kg^{-1}
1.05	31.6	310	9.8
2.1	53.7	540	10.1
4.2	64.6	810	12.5

Data adapted from Rose and Biernacka (1999).

4.12 Consider the data given below for cotton (*Gossypium hirsutum* L.) grown on Decatur silt loam (clayey, thermic, kaolinitic Rhodic Paleudults) in Alabama.
   a. Estimate plant N concentration N_c, plant P concentration P_c, and plant K concentration K_c for each sampling date.
   b. Plot dry matter Y, N_u, and N_c vs. time t on linear graph paper.
   c. Assume the parameters $t_i = 20\,\text{wk}$, $\mu = 26\,\text{wk}$, $\sqrt{2}\sigma = 8\,\text{wk}$, $c = 0.1\,\text{wk}^{-1}$, $k = 5$. Show that dimensionless time x and the growth quantifier ΔQ are given by

$$x = \frac{t - \mu}{\sqrt{2}\sigma} + \frac{\sqrt{2}\sigma c}{2} = \frac{t - 26}{8} + 0.4 = \frac{t - 22.8}{8}$$

$$\Delta Q = 0.756\{2.75(\text{erf } x + 0.3794) - 2.821[\exp(-x^2) - 0.8847]\}$$

   d. Calculate the growth quantifier ΔQ vs. t ($20 \le t \le 40\,\text{wk}$) from the model and plot the simulation curve on (b).
   e. Estimate Y from (ΔQ using the equation

$$Y = \frac{5.6}{3.383}\Delta Q = 1.65\Delta Q$$

   f. Plot the phase equation, N_u vs. Y from the model

$$N_u = \frac{250\,Y}{6.0 + Y}$$

   along with the data on linear graph paper.
   g. Plot the phase equation, P_u vs. Y, from the model

$$P_u = \frac{26\,Y}{6.0 + Y}$$

   along with the data on linear graph paper.
   h. Plot the phase equation K_u vs. Y from the model

$$K_u = \frac{190\,Y}{6.0 + Y}$$

   along with the data on linear graph paper.

i. Discuss the results.

Accumulation of dry matter and plant nutrients
by cotton in Alabama

t	Y	N_u	P_u	K_u
wk	Mg ha^{-1}	kg ha^{-1}	kg ha^{-1}	kg ha^{-1}
21.6	0.35	16	0.15	9
23.7	1.0	40	3.4	20
25.9	2.7	80	7.7	60
28.0	4.6	120	7.4	90
30.3	5.7	130	15	100
32.4	6.6	140	15	90
34.7	7.8	135	16	110

Data adapted from Mullins and Burmester (1990).

4.13 Consider the data given below for cotton (*Gossypium hirsutum* L.)
grown on Avondale clay loam [fine-loamy, mixed (clacerous),
hyperthermic, Anthropic Torrifluvent] in Arizona.
a. Plot dry matter of total above ground plant Y vs. CO_2 concentration on linear paper.
b. Plot response of the logistic model

$$\text{Dry: } Y = \frac{1800}{1 + \exp(2.03 - 0.0057 CO_2)}$$

$$\text{Wet: } Y = \frac{2600}{1 + \exp(2.03 - 0.0057 CO_2)}$$

on (a).
c. Plot seed cotton X vs. whole plant Y on linear graph paper.
d. Obtain a linear relationship between seed cotton vs. whole plant by
linear regression.
e. Obtain a restricted linear equation (constrained to pass through the
0, 0 intercept) from

$$\hat{X} = \frac{\sum_{i=1}^{6} X_i Y_i}{\sum_{i=1}^{6} Y_i^2} \, Y = 0.393 \, Y$$

Plot $\hat{X}$ vs. Y on (c).
f. Does it appear reasonable to assume that seed cotton is a constant
fraction of whole plant?
g. Estimate the concentration of CO_2 which produces 50% of projected maximum Y.

Response of cotton to CO_2 concentrations at Phoenix, AZ

Irrigation	CO_2 μmol mol^{-1}	Whole plant g m^{-2}	Seed cotton g m^{-2}
Dry	350	918	356
	500	1219	493
	650	1515	654
Wet	350	1253	482
	500	1828	697
	650	2199	842

Data adapted from Kimball and Mauney (1993).

4.14 For the expanded growth model, the incremental growth quantifier ΔQ_i is given by

$$\Delta Q_i = \exp(\sqrt{2}\sigma c x_i)\left\{(1 - kx_i)(\text{erf } x_{i+1} - \text{erf } x_i)\right.$$
$$\left. - \frac{k}{\sqrt{\pi}} [\exp(-x_{i+1}^2) - \exp(-x_i^2)]\right\}$$

The seasonal growth quantifier Q_t is given by

$$Q_t = \sum_{-\infty}^{+\infty} \Delta Q_i = \sum_{-\infty}^{+\infty} \exp(\sqrt{2}\sigma c x_i)$$
$$\left\{(1 - kx_i)(\text{erf } x_{i+1} - \text{erf } x_i) - \frac{k}{\sqrt{\pi}} [\exp(-x_{i+1}^2) - \exp(-x_i^2)]\right\}$$

Assume the parameters $\mu = 26$ wk, $\sqrt{2}\sigma = 8$ wk, $c = 0.2$ wk^{-1}, $k = 5$.

a. Confirm the equations

$$x = \frac{t - \mu}{\sqrt{2}\sigma} + \frac{\sqrt{2}\sigma c}{2} = \frac{t - 26.0}{8} + \frac{(8)(0.2)}{2} = \frac{t - 19.6}{8}$$

$$\Delta Q_i = \exp\left(\sqrt{2}\sigma c x_i\right)\left\{(1 - kx_i)(\text{erf } x_{i+1} - \text{erf } x_i)\right.$$
$$\left. - \frac{k}{\sqrt{\pi}} [\exp(-x_{i+1}^2) - \exp(-x_i^2)]\right\}$$
$$= \exp(1.6x_i)\{(1 - 5x_i)(\text{erf } x_{i+1} - \text{erf } x_i)$$
$$- 2.821[\exp(-x_{i+1}^2) - \exp(-x_i^2)]$$

b. Use the values in the attached table to calculate ΔQ_i for harvest intervals of $\Delta t = 2$, 4, 6, 8, 12, and 16 wk starting at $t_i = 0$. Compare your results to the second attached table.

c. Calculate the seasonal growth quantifier $Q_t = \sum Q_i$ for each Δt.

d. Calculate the standardized seasonal growth quantifier $Q_t^* = Q_t \exp(0.10\Delta t)$ for each Δt.

e. Perform linear regression on Q_t^* vs. Δt to obtain the coefficients α and β for the equation

$$Q_t^* = \alpha + \beta \Delta t$$

f. Plot the data for Q_t and Q_t^* vs. Δt on linear graph paper.

g. Plot the curve on (e) from the model

$$Q_t = (\alpha + \beta \Delta t) \exp(-0.10\Delta t)$$

h. Plot the line on (e) from the model

$$Q_t^* = \alpha + \beta \Delta t$$

i. Discuss the results.

4.15 The phenomenological growth model assumes a linear intrinsic growth function. The cumulative growth quantifier Q_n is given by

$$
\begin{aligned}
Q_n &= \sum_{i=1}^{n} \left\{ (1 - kx_i)(\operatorname{erf} x_{i+1} - \operatorname{erf} x_i) \right. \\
&\quad \left. - \frac{k}{\sqrt{\pi}} [\exp(-x_{i+1}^2) - \exp(-x_i^2)] \right\} \\
&= \left(1 + \frac{k\Delta x}{2} \right) \sum_{i=1}^{n} (\operatorname{erf} x_{i+1} - \operatorname{erf} x_i) \\
&\quad - k \sum_{i=1}^{n} x_{i+1/2}(\operatorname{erf} x_{i+1} - \operatorname{erf} x_i) \\
&\quad - \frac{k}{\sqrt{\pi}} \sum_{i=1}^{n} [\exp(-x_{i+1}^2) - \exp(-x_i^2)]
\end{aligned}
$$

Use the table below for the following questions.

a. Calculate the first sum on the right from

$$\sum_{x_i=-\infty}^{n} (\operatorname{erf} x_{i+1} - \operatorname{erf} x_i) \approx \sum_{x_i=-4}^{n} (\operatorname{erf} x_{i+1} - \operatorname{erf} x_i)$$

for $-4 \le x_i \le +4$ and $\Delta x = 1$.

b. Calculate the second sum on the right from

$$\sum_{x_i=-\infty}^{n} x_{i+1/2}(\text{erf}\, x_{i+1} - \text{erf}\, x_i) \approx \sum_{x_i=-4}^{n} x_{i+1/2}(\text{erf}\, x_{i+1} - \text{erf}\, x_i)$$

c. Calculate the third sum on the right from

$$\sum_{x_i=-\infty}^{n} [\exp(-x_{i+1}^2) - \exp(-x_i^2)] \approx \sum_{x_i=-4}^{n} [\exp(-x_{i+1}^2) - \exp(-x_i^2)]$$

d. Calculate the difference from

$$\sum_{x_i=-4}^{n} x_{i+1/2}(\text{erf}\, x_{i+1} - \text{erf}\, x_i) - \frac{1}{\sqrt{\pi}} \sum_{x_i=-4}^{n} [\exp(-x_{i+1}^2) - \exp(-x_i^2)]$$

e. Does (d) appear to follow a Gaussian distribution centered about zero?
f. Show that the total sum

$$\sum_{x_i=-4}^{+4} (\text{erf}\, x_{i+1} - \text{erf}\, x_i) = 2$$

g. Show that the total sum

$$\sum_{x_i=-4}^{+4} x_{i+1/2}(\text{erf}\, x_{i+1} - \text{erf}\, x_i) = 0$$

h. Show that the total sum

$$\sum_{x_i=-4}^{+4} [\exp(-x_{i+1}^2) - \exp(-x_i^2)] = 0$$

i. Calculate Q_n for each x_n for $k = 5$.
j. Calculate $F = Q_n/Q_t$ for each x_n.
k. Plot F vs. x on probability paper. Does it appear to follow a straight line?

4.16 Evans et al. (1961) measured response of Coastal bermudagrass and Pensacola bahiagrass to applied N and irrigation on Greenville fine sandy loam (fine, kaolinitic, thermic Rhodic Kandiudults) at Thorsby, AL. Results are given in the table below.

a. Plot average dry Y for the years 1957–1959 vs. applied nitrogen N for both grasses for both irrigated and nonirrigated on linear graph paper.

b. Draw the model curves on (a) from the equations

Bermudagrass, nonirrigated: $Y = \dfrac{21.57}{1 + \exp(1.39 - 0.0078N)}$

Bermudagrass, irrigated: $Y = \dfrac{23.44}{1 + \exp(1.39 - 0.0078N)}$

Bahiagrass, nonirrigated: $Y = \dfrac{21.49}{1 + \exp(1.57 - 0.0078N)}$

Bahiagrass, irrigated: $Y = \dfrac{22.73}{1 + \exp(1.57 - 0.0078N)}$

c. Construct the scatter plot (estimated vs. measured yield) for the four cases on linear graph paper.
d. Discuss the results. Does the logistic model describe the data well? Would you judge there to be a significant difference between irrigated and nonirrigated plots? How do the two cultivars compare?

Seasonal dry matter yield of coastal bermudagrass and pensacola bahiagrass grown on Greenville fine sandy loam at Thorsby, AL

			Applied N, kg ha^{-1}			
			0	168	336	672
Species	Irrigation	Year		Mg ha^{-1}		
Bermuda	No	1957	3.74	12.52	17.41	22.32
		1958	4.70	11.38	17.44	21.56
		1959	3.40	9.04	13.36	20.44
	Yes	1957	3.49	11.70	18.32	21.08
		1958	4.33	10.75	17.89	22.61
		1959	4.76	11.94	19.42	24.43
Bahia	No	1957	4.56	11.70	16.39	20.32
		1958	3.63	10.39	16.85	22.23
		1959	2.67	8.19	14.14	19.62
	Yes	1957	4.31	10.35	16.81	23.26
		1958	3.24	9.25	15.69	22.33
		1959	3.82	9.52	16.33	22.43

Data adapted from Evans et al. (1961).

4.17 Evers (1984) measured the response of coastal bermudagrass and pensacola bahiagrass to applied N and irrigation on Crowley sandy loam (fine, montmorillonitic, thermic Typic Albaqualfs) soil at Eagle Lake,

TX. Results are given in the table below. Wilson (1995) has analyzed these data in detail.

a. Plot dry matter Y, plant N uptake N_u, and plant N concentration N_c vs. applied N, N, on linear graph paper for each year and irrigation treatment.

b. Draw the model curves on (a) from the equations

$$Y = \frac{A}{1 + \exp(0.57 - 0.0072N)}$$

$$N_u = \frac{A_n}{1 + \exp(1.07 - 0.0072N)}$$

$$N_c = N_{cm} \left[\frac{1 + \exp(0.57 - 0.0072N)}{1 + \exp(1.07 - 0.0072N)} \right]$$

where the parameters for bermudagrass plots are given by the values $A(1979) = 15.99$, $A(1980) = 9.89$ Mg ha^{-1}; $A_n(1979) = 311$, $A_n(1980) = 227$ kg ha^{-1}; N_{cm} (1979) $= 19.4$, $N_{cm}(1980) = 23.0$ g kg^{-1}; and for bahiagrass plots are given by the values: $A(1979) = 12.42$, $A(1980) = 7.51$ Mg ha^{-1}; $A_n(1979) = 237$, $A_n(1980) = 191$ kg ha^{-1}; N_{cm} (1979) $= 19.1$, $N_{cm}(1980) = 25.5$ g kg^{-1}.

c. Plot the phase relations Y vs. N_u and N_c vs. N_u on linear graph paper.

d. Plot the model curves on (c) from the equations

$$Y = \frac{Y_m N_u}{K_n + N_u}$$

$$N_c = \frac{K_n}{Y_m} + \frac{1}{Y_m} N_u$$

where for bermudagrass Y_m (1979) $= 40.6$ Mg ha^{-1}, $K_n(1979) = 479$ kg ha^{-1}, $Y_m(1980) = 25.1$ Mg ha^{-1}, $K_n(1980) = 351$ kg ha^{-1}; and for bahiagrass Y_m (1979) $= 31.6$ Mg ha^{-1}, $K_n(1979) = 366$ kg ha^{-1}, Y_m (1980) $= 19.1$ Mg ha^{-1}, $K_n(1980) = 295$ kg ha^{-1}.

e. Discuss agreement between the model and data for both years.

Seasonal dry matter yield, plant N removal, and plant N concentration
for coastal bermudagrass and pensacola bahiagrass at Eagle Lake, TX

Grass	Year	Applied N kg ha^{-1}	Dry matter Mg ha^{-1}	N removal kg ha^{-1}	N concentration g kg^{-1}
Bermuda	1979	0	4.47	58	13.0
		84	6.98	104	14.9
		168	10.00	156	15.5
		252	12.91	225	17.4
		336	14.23	258	18.1
	1980	0	3.90	62	16.0
		84	5.47	102	18.7
		168	6.33	116	18.4
		252	7.47	155	20.7
		336	8.12	178	21.9
Bahia	1979	0	4.38	55	12.6
		84	6.02	88	14.6
		168	7.71	123	16.0
		252	10.10	173	17.1
		336	10.40	186	17.9
	1980	0	2.95	65	22.0
		84	3.99	81	20.2
		168	5.17	108	20.9
		252	5.30	120	22.7
		336	6.32	150	23.7

Data adapted from Evers (1984).

4.18 Jeffers (1955) conducted field experiments with coastal bermudagrass
and pensacola bahiagrass on Red Bay sandy loam (fine, siliceous,
thermic Rhodic Paleudults) at Jay, FL. Plots overseeded with clover
received no applied N. Average harvest intervals were 5.8 wk for 1954
and 7.7 wk for 1955. Data are given in the table below. Analysis of
these data is described in greater detail by Overman et al. (1992).
  a. Plot dry matter yield vs. applied N for the two grasses and two
     years on linear graph paper.
  b. Draw the model curves on (a) from the equations given.

Bermudagrass 1954: $Y = \dfrac{15.07}{1 + \exp(1.71 - 0.0065N)}$

1955: $Y = \dfrac{19.22}{1 + \exp(1.71 - 0.0065N)}$

$$\text{Bahiagrass } 1954: \quad Y = \frac{19.39}{1 + \exp(1.71 - 0.0065N)}$$

$$1955: \quad Y = \frac{22.68}{1 + \exp(1.71 - 0.0065N)}$$

c. Discuss the results. Compare values of the A parameter between years and offer at least two possible explanations for these differences.

d. Estimate the equivalent N supplied by the clover for each grass and each year.

Seasonal dry matter yield for coastal bermudagrass and pensacola bahiagrass at Jay, FL

Grass	Year	Applied N kg ha^{-1}	Yield Mg ha^{-1}
Bermuda	1954	38	1.90
		76	2.84
		140	5.60
		280	8.80
		560	12.68
		clover	4.14
	1955	56	3.79
		112	5.22
		224	8.96
		448	14.16
		896	19.20
		clover	5.08
Bahia	1954	38	2.91
		76	4.30
		140	6.68
		280	11.13
		560	16.35
		clover	4.50
	1955	56	4.28
		112	5.71
		224	10.15
		448	16.73
		896	23.00
		clover	5.76

Data adapted from Jeffers (1955).

4.19 Doss et al. (1966) studied response of bermudagrass to applied N and harvest interval at Thorsby, AL. Data are given in the table below.
  a. Plot dry matter Y, plant N uptake N_u, and plant N concentration N_c vs. applied N, N, on linear graph paper.
  b. Draw the model curves on (a) from the equations

$$3.0 \text{ wk: } Y = \frac{17.42}{1 + \exp(1.27 - 0.0067N)}$$

$$N_u = \frac{569}{1 + \exp(2.02 - 0.0067N)}$$

$$N_c = 32.7 \left[\frac{1 + \exp(1.27 - 0.0067N)}{1 + \exp(2.02 - 0.0067N)} \right]$$

$$4.5 \text{ wk: } Y = \frac{19.75}{1 + \exp(1.27 - 0.0067N)}$$

$$N_u = \frac{544}{1 + \exp(2.02 - 0.0067N)}$$

$$N_c = 27.5 \left[\frac{1 + \exp(1.27 - 0.0067N)}{1 + \exp(2.02 - 0.0067N)} \right]$$

  c. Discuss the dependence of Y, N_u, and N_c on applied N.
  d. Plot the phase relations Y vs. N_u and N_c vs. N_u on linear graph paper.
  e. Plot the model curves on (d) from the equations

$$3.0 \text{ wk: } Y = \frac{33.0N_u}{509 + N_u}$$

$$N_c = 15.4 + 0.0303N_u$$

$$4.5 \text{ wk: } Y = \frac{37.4N_u}{487 + N_u}$$

$$N_c = 13.0 + 0.0267N_u$$

  f. Construct scatter plots (estimated vs. measured yield and estimated vs. measured plant N uptake) for the four cases on linear graph paper. Discuss agreement between the model and data.
4.20 Huneycutt et al. (1988) measured the response of bermudagrass to applied N and irrigation at Fayetteville, AR.
  a. Plot dry matter Y, plant N uptake N_u, and plant N concentration N_c vs. applied N, N, on linear graph paper for each year and irrigation treatment.

Seasonal dry matter yield, plant N removal, and plant N
concentration for bermudagrass harvested at 3.0 and 4.5 wk
intervals and grown at Thorsby, AL

Applied N kg ha^{-1}	Dry matter Mg ha^{-1}		N removal kg ha^{-1}		N concentration g kg^{-1}	
	3.0 wk	4.5 wk	3.0 wk	4.5 wk	3.0 wk	4.5 wk
0	2.95	3.60	53	55	18.1	15.4
224	9.70	11.95	230	220	23.7	18.4
448	14.70	17.20	406	387	27.6	22.5
672	16.70	18.50	501	466	30.0	25.2
1344	17.50	19.95	560	549	32.0	27.5
2016	17.60	19.20	597	568	33.9	29.6

Data adapted from Doss et al. (1966).

b. Draw the model curves on (a) from the equations

$$Y = \frac{A}{1 + \exp(1.50 - 0.0084N)}$$

$$N_u = \frac{A_n}{1 + \exp(2.04 - 0.0084N)}$$

$$N_c = N_{cm}\left[\frac{1 + \exp(1.50 - 0.0084N)}{1 + \exp(2.04 - 0.0084N)}\right]$$

where the parameters for nonirrigated plots are given by the values:
$A(1983) = 17.90$, $A(1984) = 17.37$, $A(1985) = 19.61$ Mg ha^{-1};
$A_n(1983) = 409$, $A_n(1984) = 414$, $A_n(1985) = 459$ kg ha^{-1};
$N_{cm}(1983) = 22.8$, $N_{cm}(1984) = 23.8$, $N_{cm}(1985) = 23.4$ g kg^{-1};
and for irrigated plots are given by the values: $A(1983) = 24.70$,
$A(1984) = 24.60$, $A(1985) = 22.58$ Mg ha^{-1}; $A_n(1983) = 554$,
$A_n(1984) = 524$, $A_n(1985) = 492$ kg ha^{-1}; $N_{cm}(1983) = 22.4$,
$N_{cm}(1984) = 21.3$, $N_{cm}(1985) = 21.8$ g kg^{-1}.

c. Plot the phase relations Y vs. N_u and N_c vs. N_u on linear graph
paper.

d. Plot the model curves on (c) from the equations

Nonirrigated   1983: $$Y = \frac{42.9N_u}{571 + N_u}$$

$$N_c = 13.3 + 0.0233N_u$$

$$1984: \quad Y = \frac{41.6N_u}{578 + N_u}$$

$$N_c = 13.9 + 0.0240N_u$$

$$1985: \quad Y = \frac{47.0N_u}{641 + N_u}$$

$$N_c = 13.6 + 0.0213N_u$$

Irrigated 1983: $\quad Y = \dfrac{59.2N_u}{774 + N_u}$

$$N_c = 13.1 + 0.0169N_u$$

$$1984: \quad Y = \frac{59.0N_u}{731 + N_u}$$

$$N_c = 12.4 + 0.0170N_u$$

$$1985: \quad Y = \frac{54.1N_u}{687 + N_u}$$

$$N_c = 12.7 + 0.0185N_u$$

e. Construct scatter plots (estimated vs. measured yield and estimated vs. measured plant N uptake) for the four cases on linear graph paper. Discuss agreement between the model and data.

Seasonal dry matter yield, plant N uptake, and plant N concentration for bermudagrass grown at Fayetteville, AR

Component	Year	\multicolumn{7}{c}{Applied nitrogen, kg ha^{-1}}
		0	112	224	336	448	560	672
		\multicolumn{7}{c}{Nonirrigated}
Dry matter	1983	1.95	6.88	11.52	13.88	16.39	16.70	17.71
Mg ha^{-1}	1984	2.35	8.43	11.59	12.71	16.50	16.48	16.01
	1985	1.70	7.20	9.37	15.42	19.70	18.70	19.41
		\multicolumn{7}{c}{Irrigated}
	1983	4.53	9.12	14.50	20.80	21.90	23.63	23.87
	1984	4.17	11.52	14.17	18.31	22.42	23.67	24.34
	1985	1.75	8.32	12.55	18.56	20.11	21.36	23.18
		\multicolumn{7}{c}{Nonirrigated}
N removal	1983	27	109	210	262	354	358	419
kg ha^{-1}	1984	29	136	210	266	370	393	379
	1985	21	105	178	294	416	407	484

continued

Component	Year	Applied nitrogen, kg ha^{-1}						
		0	112	224	336	448	560	672
		Irrigated						
	1983	67	156	278	379	466	537	512
	1984	61	186	252	366	402	451	561
	1985	25	129	229	333	376	458	523
		Nonirrigated						
N concentration	1983	13.8	15.8	18.2	18.9	21.6	21.4	23.7
g kg^{-1}	1984	12.2	16.2	18.0	21.0	22.4	23.8	23.7
	1985	12.5	14.6	19.0	19.0	21.1	21.8	25.0
		Irrigated						
	1983	14.7	17.1	19.2	18.2	21.3	22.7	21.4
	1984	14.7	16.2	17.8	20.0	17.9	19.0	23.0
	1985	14.1	15.5	18.2	17.9	18.7	21.4	22.6

Data adapted from Huneycutt et al. (1988).

4.21 Huneycutt et al. (1988) measured the response of tall fescue to applied N and irrigation at Fayetteville, AR.

   a. Plot dry matter Y, plant N uptake N_u, and plant N concentration N_c vs. applied N, N, on linear graph paper for each year and irrigation treatment.

   b. Draw the model curves on (a) from the equations

$$Y = \frac{A}{1 + \exp(0.92 - 0.0081N)}$$

$$N_u = \frac{A_n}{1 + \exp(1.47 - 0.0081N)}$$

$$N_c = N_{cm}\left[\frac{1 + \exp(0.92 - 0.0081N)}{1 + \exp(1.47 - 0.0081N)}\right]$$

where the parameters for nonirrigated plots are given by the values: $A(1981-1982) = 12.08$, $A(1982-1983) = 8.01$, $A(1983-1984) = 5.71$ Mg ha^{-1}; $A_n(1981-1982) = 358$, $A_n(1982-1983) = 218$, $A_n(1983-1984) = 170$ kg ha^{-1}; N_{cm} (1981–1982) = 29.6, N_{cm} (1982–1983) = 27.2, N_{cm} (1983–1984) = 29.7 g kg^{-1}; and for irrigated plots are given by the values: $A(1981-1982) = 15.63$, $A(1982-1983) = 13.22$, $A(1983-1984) = 16.81$ Mg ha^{-1}; $A_n(1981-1982) = 444$, $A_n(1982-1983) = 357$, $A_n(1983-1984) = 439$ kg ha^{-1};

N_{cm} (1981–1982) = 28.4, N_{cm} (1982–1983) = 27.0, N_{cm} (1983–1984) = 26.1 g kg^{-1}.

c. Plot the phase relations Y vs. N_u and N_c vs. N_u on linear graph paper.

d. Plot the model curves on (c) from the equations

Nonirrigated 1981–1982: $Y = \dfrac{28.6N_u}{488 + N_u}$

$N_c = 17.1 + 0.0350N_u$

1982–1983: $Y = \dfrac{18.9N_u}{297 + N_u}$

$N_c = 15.7 + 0.0528N_u$

1983–1984: $Y = \dfrac{13.5N_u}{231 + N_u}$

$N_c = 17.1 + 0.0741N_u$

Irrigated 1981–1982: $Y = \dfrac{36.9N_u}{605 + N_u}$

$N_c = 16.4 + 0.0271N_u$

1982–1983: $Y = \dfrac{31.2N_u}{487 + N_u}$

$N_c = 15.6 + 0.0320N_u$

1983–1984: $Y = \dfrac{39.7N_u}{599 + N_u}$

$N_c = 15.1 + 0.0252N_u$

e. Construct scatter plots (estimated vs. measured yield and estimated vs. measured plant N uptake) for the four cases on linear graph paper. Discuss agreement between the model and data.

Seasonal dry matter yield, plant N uptake, and plant N concentration for tall fescue grown at Fayetteville, AR

		Applied nitrogen, kg ha^{-1}						
Component	Year	0	112	224	336	448	560	672
		Nonirrigated						
Dry matter	1981–1982	1.95	6.88	11.52	13.88	16.39	16.70	17.71
Mg ha^{-1}	1982–1983	2.35	8.43	11.59	12.71	16.50	16.48	16.01
	1983–1984	1.70	7.20	9.37	15.42	19.70	18.70	19.41

continued

Component	Year	Applied nitrogen, kg ha^{-1}						
		0	112	224	336	448	560	672
				Irrigated				
	1981–1982	4.53	9.12	14.50	20.80	21.90	23.63	23.87
	1982–1983	4.17	11.52	14.17	18.31	22.42	23.67	24.34
	1983–1984	1.75	8.32	12.55	18.56	20.11	21.36	23.18
				Nonirrigated				
N removal	1981–1982	27	109	210	262	354	358	419
kg ha^{-1}	1982–1983	29	136	210	266	370	393	379
	1983–1984	21	105	178	294	416	407	484
				Irrigated				
	1981–1982	67	156	278	379	466	537	512
	1982–1983	61	186	252	366	402	451	561
	1983–1984	25	129	229	333	376	458	523
				Nonirrigated				
N concen-	1981–1982	13.8	15.8	18.2	18.9	21.6	21.4	23.7
tration								
g kg^{-1}	1982-1983	12.2	16.2	18.0	21.0	22.4	23.8	23.7
	1983–1984	12.5	14.6	19.0	19.0	21.1	21.8	25.0
				Irrigated				
	1981–1982	14.7	17.1	19.2	18.2	21.3	22.7	21.4
	1982–1983	14.7	16.2	17.8	20.0	17.9	19.0	23.0
	1983–1984	14.1	15.5	18.2	17.9	18.7	21.4	22.6

Data adapted from Huneycutt et al. (1988).

4.22 Morrison et al. (1980) conducted an extensive study at 20 sites in England on response of rye grass to applied N. Average data for the 20 sites are given in the table below.
   a. Plot dry matter Y, plant N uptake N_u, and plant N concentration N_c vs. applied N, N, on linear graph paper.
   b. Draw the model curves on (a) from the equations

$$Y = \frac{12.10}{1 + \exp(1.06 - 0.0074N)}$$

$$N_u = \frac{430}{1 + \exp(1.78 - 0.0074N)}$$

$$N_c = 35.5 \left[\frac{1 + \exp(1.06 - 0.0074N)}{1 + \exp(1.78 - 0.0074N)} \right]$$

c. Plot the phase relations Y vs. N_u and N_c vs. N_u on linear graph paper.

d. Plot the model curves on (c) from the phase equations

$$Y = \frac{Y_m N_u}{K_n + N_u} = \frac{23.6 N_u}{408 + N_u}$$

$$N_c = \frac{K_n}{Y_m} + \frac{1}{Y_m} N_u = 17.3 + 0.0424 N_u$$

e. Use relations developed in the text to show that $Y_m = 23.6$ Mg ha^{-1} and $K_n = 408$ kg ha^{-1}.

f. Discuss agreement between the model and data.
   Note: These data have been analyzed in greater detail by Wilson (1995).

Average seasonal dry matter yield, plant N removal, and plant N concentration for ryegrass grown in England

Applied N kg ha^{-1}	Dry matter Mg ha^{-1}	Plant N uptake kg ha^{-1}	Plant N concentration g kg^{-1}
0	2.67 ± 1.63	60 ± 41	21.4 ± 3.2
150	6.45 ± 1.81	153 ± 52	23.4 ± 2.3
300	9.63 ± 1.81	264 ± 57	27.4 ± 2.1
450	11.25 ± 2.01	354 ± 68	31.5 ± 2.4
600	11.61 ± 2.08	399 ± 67	34.5 ± 2.3
750	11.39 ± 2.09	421 ± 74	37.0 ± 2.1

Data adapted from Morrison et al. (1980).

4.23 Walker and Morey (1962) conducted a factorial experiment on response of winter rye to applied N, P, and K at Tifton, GA. Data are given in the table below.

a. Plot dry matter Y, plant N uptake N_u, and plant N concentration N_c vs. applied N, N, on linear graph paper for applied P and K of 40 and 74 kg ha^{-1}, respectively.

b. Draw the model curves on (a) from the equations

$$Y = \frac{5.43}{[1 + \exp(1.36 - 0.0225N)][1 + \exp(-0.16 - 0.0464P)]}$$
$$[1 + \exp(-0.91 - 0.0201K)]$$

$$N_u = \frac{260}{[1 + \exp(1.93 - 0.0225N)][1 + \exp(-0.16 - 0.0464P)]}$$
$$[1 + \exp(-0.91 - 0.0201K)]$$

$$N_c = 47.9 \left[\frac{1 + \exp(1.36 - 0.0225N)}{1 + \exp(1.93 - 0.0225N)} \right]$$

c. Plot dry matter Y, plant P uptake P_u, and plant P concentration P_c vs. applied P, P, on linear graph paper for applied N and K of 135 and 74 kg ha^{-1}, respectively.

d. Draw the model curves on (c) from the equations

$$Y = \frac{5.43}{[1 + \exp(1.36 - 0.0225N)][1 + \exp(-0.16 - 0.0464P)]}$$
$$[1 + \exp(-0.91 - 0.0201K)]$$

$$P_u = \frac{34}{[1 + \exp(1.36 - 0.0225N)][1 + \exp(-0.14 - 0.0464P)]}$$
$$[1 + \exp(-0.91 - 0.0201K)]$$

$$P_c = 6.26 \left[\frac{1 + \exp(-0.16 - 0.0464P)}{1 + \exp(-0.14 - 0.0464P)} \right]$$

e. Plot dry matter Y, plant K uptake K_u, and plant K concentration K_c vs. applied K, K, on linear graph paper for applied N and P of 135 and 40 kg ha^{-1}, respectively.

f. Draw the model curves on (e) from the equations

$$Y = \frac{5.43}{[1 + \exp(1.36 - 0.0225N)][1 + \exp(-0.16 - 0.0464P)]}$$
$$[1 + \exp(-0.91 - 0.0201K)]$$

$$K_u = \frac{230}{[1 + \exp(1.36 - 0.0225N)][1 + \exp(-0.16 - 0.0464P)]}$$
$$[1 + \exp(0.46 - 0.0201K)]$$

$$K_c = 42.4 \left[\frac{1 + \exp(-0.91 - 0.0201K)}{1 + \exp(0.46 - 0.0201K)} \right]$$

g. Plot the phase relations Y vs. N_u and N_c vs. N_u on linear graph paper for P and K of 40 and 74 kg ha^{-1}, respectively.

h. Plot the model curves on (g) from the equations

$$Y = \frac{10.1N_u}{273 + N_u}$$

$$N_c = 27.1 + 0.0990N_u$$

    i. Plot the phase relations Y vs. K_u and K_c vs. K_u on linear graph paper for N and P of 135 and 40 kg ha^{-1}, respectively.

    j. Plot the model curves on (i) from the equations

    k. Construct scatter plots (estimated vs. measured) for dry matter, plant N removal, and plant K removal.

    m. Discuss agreement between data and the model.

Seasonal dry matter yield, removal of plant N, P, and K, and concentration of plant N, P, and K for winter rye at Tifton, GA

N	P	K	Y	N_u	P_u	K_u	N_c	P_c	K_c
kg ha^{-1}			Mg ha^{-1}	kg ha^{-1}			g kg^{-1}		
0	40	74	0.55	—	—	—	—	—	—
45	40	74	1.81	58	12.1	66	32.3	3.69	36.4
90	40	74	3.01	112	19.6	108	37.3	6.51	36.6
135	40	74	3.75	160	24.4	136	42.6	6.51	36.4
180	40	74	4.01	175	26.8	146	43.7	6.69	36.4
225	40	74	4.55	216	28.6	148	47.5	6.29	32.5
135	0	74	2.09	92	13.0	75	44.1	6.21	35.8
135	20	74	3.08	134	17.0	118	43.5	5.51	38.3
135	40	74	3.36	133	21.1	119	39.5	6.29	35.4
135	60	74	3.74	146	24.3	141	39.1	6.51	37.8
135	80	74	4.00	148	21.7	134	37.0	5.42	33.4
135	100	74	3.97	—	—	—	—	—	—
135	40	0	2.87	112	27.6	62	39.0	9.61	21.7
135	40	37	3.18	116	15.3	90	36.6	4.81	28.3
135	40	74	3.74	140	24.3	133	37.3	6.51	35.5
135	40	111	3.56	134	18.8	130	37.5	5.29	36.4
135	40	148	3.96	160	24.2	150	40.4	6.12	37.8
135	40	185	4.11	162	31.2	159	39.4	7.60	38.8

Data adapted from Walker and Morey (1962).

4.24 Rhoads et al. (1997) studied response of pensacola bahiagrass to applied N, P, and K on Dothan loamy fine sand (fine-loamy, siliceous, thermic Plinthic Kandiudults) at Quincy, FL. Results from the incomplete factorial experiments are given in Table 1 below.

a. Average the yields Y in Table 1 at each applied nitrogen N and applied potassium K for applied phosphorus P of 84 and $168 \, kg \, ha^{-1}$ to obtain the values in Table 2.

b. Use the values of the parameter A_n listed in Table 3 to estimate the nitrogen parameters for the logistic model

$$\ln\left(\frac{A_n}{Y} - 1\right) = b_n - c_n N$$

c. Estimate the potassium parameters for the logistic model from the equation

$$\ln\left(\frac{15.00}{A_n} - 1\right) = b_k - c_k K = -0.56 - 0.0171K \quad r = -0.9972$$

d. Calculate standardized yields Y^* over N and K given in Table 4 from the equation

$$Y^* = Y\,[1 + \exp(0.27 - 0.0072N)][1 + \exp(-0.56 - 0.0171K)]$$

e. Use the values for $N = 0$ and $K = 224 \, kg \, ha^{-1}$ in Table 4 to estimate phosphorus parameters for the logistic model

$$\ln\left(\frac{15.50}{Y^*} - 1\right) = b_p - c_p P = -0.69 - 0.0223P \quad r = -1.0000$$

Since this equation underestimates yields at low P levels, the intercept parameter is adjusted to $b_p = -1.4$.

f. Calculate reduced yields listed in Table 5 from the triple logistic equation

$$\frac{Y}{A} = \frac{1}{[1 + \exp(0.27 - 0.0072N)][1 + \exp(-1.4 - 0.223P)]} \\ [1 + \exp(-0.56 - 0.0171K)]$$

g. Calculate the weighted-value of parameter A from the equation

$$A = \frac{\displaystyle\sum \frac{Y}{[1 + \exp(0.27 - 0.0072N)][1 + \exp(-1.4 - 0.223P)]} }{[1 + \exp(-0.56 - 0.0171K)]}}{\displaystyle\sum \frac{1}{[1 + \exp(0.27 - 0.0072N)][1 + \exp(-1.4 - 0.223P)]} }{[1 + \exp(-0.56 - 0.0171K)]}}$$

$$= \frac{130.46}{8.608} = 15.16 \, Mg \, ha^{-1}$$

h. Calculate the estimated dry matter yields given in Table 6 from the triple logistic equation

$$\hat{Y} = \frac{15.16}{[1 + \exp(0.27 - 0.0072N)][1 + \exp(-1.4 - 0.223P)]}{[1 + \exp(-0.56 - 0.0171K)]}$$

i. Plot estimated vs. measured Y for the 27 combinations of N, P, and K on linear graph paper.

j. Show that the correlation between measured and estimated Y in the scatter diagram (i) is given by

$$\hat{Y} = -0.09 + 1.0025\,Y \qquad r = 0.9784$$

k. Discuss the results. This procedure is obviously tedious. Would nutrient uptake data for N, P, and K help in the estimation process?

Table 1 Yield response of pensacola bahiagrass to applied N, P, and K at Quincy, FL

Applied N kg ha^{-1}	Applied P kg ha^{-1}	Dry matter Mg ha^{-1} Applied K, kg ha^{-1}		
		0	112	224
0	0	4.07	5.11	4.43
0	84	4.62	5.98	6.15
0	168	4.60	5.47	6.55
168	0	6.55	7.78	8.42
168	84	7.15	9.25	10.89
168	168	6.90	8.84	10.21
336	0	8.25	10.66	10.06
336	84	8.54	12.50	13.73
336	168	8.10	12.28	13.19

Data adapted from Rhoads et al. (1997).

Table 2 Yield response of pensacola bahiagrass to applied N and K at Quincy, FL

Applied N kg ha^{-1}	Dry matter Mg ha^{-1} Applied K, kg ha^{-1}		
	0	112	224
0	4.61	5.72	6.35
168	7.02	9.05	10.55
336	8.32	12.39	13.46

Yield values are averaged over P = 84 and 168 kg ha^{-1}.

Table 3 Linearized yield response of bahiagrass to applied N and K at Quincy, FL

Applied N kg ha^{-1}	$A_n/Y - 1$ K, kg ha^{-1}			
	0	112	224	avg
0	1.01	1.45	1.33	1.26
168	0.318	0.547	0.403	0.423
336	0.112	0.130	0.100	0.114
A_n, Mg ha^{-1}	9.25	14.00	14.80	—
b_n	−0.012	0.45	0.32	0.27
c_n, ha kg^{-1}	0.0065	0.0072	0.0077	0.0072
r	−0.9996	−0.9939	−0.9990	−0.9986

Table 4 Standardized yield response of pensacola bahiagrass to applied N, P, and K at Quincy, FL

		Y^* Mg ha^{-1}		
Applied N kg ha^{-1}	Applied P kg ha^{-1}	Applied K, kg ha^{-1}		
		0	112	224
0	0	14.77	12.80	10.36
0	84	16.77	14.98	14.38
0	168	16.70	13.70	15.32
168	0	14.31	11.73	11.86
168	84	15.62	13.95	15.33
168	168	15.08	13.33	14.38
336	0	14.47	12.90	11.37
336	84	14.98	15.13	15.52
336	168	14.21	14.87	14.91

Table 5 Reduced yield response of pensacola bahiagrass to applied N, P, and K at Quincy, FL

		Y/A Mg ha^{-1}		
Applied N kg ha^{-1}	Applied P kg ha^{-1}	Applied K, kg ha^{-1}		
		0	112	224
0	0	0.221	0.320	0.343
0	84	0.265	0.385	0.412
0	168	0.274	0.397	0.425
168	0	0.367	0.532	0.570
168	84	0.441	0.639	0.684
168	168	0.455	0.659	0.706
336	0	0.457	0.663	0.710
336	84	0.549	0.796	0.852
336	168	0.567	0.821	0.880

Table 6 Estimated yield response of pensacola bahiagrass to applied N, P, and K at Quincy, FL

Applied N kg ha^{-1}	Applied P kg ha^{-1}	Dry matter Mg ha^{-1} Applied K, kg ha^{-1} 0	112	224
0	0	3.35	4.85	5.20
0	84	4.02	5.84	6.25
0	168	4.15	6.02	6.44
168	0	5.56	8.06	8.64
168	84	6.69	9.69	10.37
168	168	6.90	9.99	10.70
336	0	6.93	10.05	10.76
336	84	8.32	12.07	12.92
336	168	8.60	12.45	13.34

4.25 Adams et al. (1966) and Carreker et al. (1977) reported on response of coastal bermudagrass to applied N, P, and K on Cecil sandy loam (fine, kaolinitic, thermic Typic Kanhapludults)at Watkinsville, GA. Data from this study are reported in the tables below. Overman et al. (1991) and Overman and Wilkinson (1995) developed a multiple logistic model for describing response of dry matter and plant nutrient uptake for these data.

a. Plot dry matter yield Y, plant N uptake N_u, and plant N concentration N_c vs. applied nitrogen N for applied phosphorus of $P = 100$ kg ha^{-1} and applied potassium of $K = 0$ and $K = 188$ kg ha^{-1} on linear graph paper.

b. Draw the model curves on (a) from the equations

$$Y = \frac{15.54}{[1 + \exp(0.462 - 0.0122N)][1 + \exp(-1.58 - 0.0410P)]}{[1 + \exp(-1.58 - 0.0212K)]}$$

$$N_u = \frac{380}{[1 + \exp(1.25 - 0.0122N)][1 + \exp(-1.58 - 0.0410P)]}{[1 + \exp(-1.58 - 0.0212K)]}$$

$$N_c = 24.4\left[\frac{1 + \exp(0.462 - 0.0122N)}{1 + \exp(1.25 - 0.0122N)}\right]$$

for the same P and K as in (a).

c. Construct the phase plot of Y and N_c vs. N_u for the same P and K as in (a).

d. Draw the model curves on (c) from the equations

$$P = 100, K = 0 \, \mathrm{kg \, ha^{-1}}: \qquad Y = \frac{23.5 N_u}{261 + N_u}$$

$$N_c = 11.1 + 0.0425 N_u$$

$$P = 100, K = 188 \, \mathrm{kg \, ha^{-1}}: \qquad Y = \frac{28.3 N_u}{313 + N_u}$$

$$N_c = 11.1 + 0.0353 N_u$$

on linear graph paper.

e. Discuss the results. Comment in particular on agreement between the model and data.

Table 1 Yield response of coastal bermudagrass to applied N, P, and K at Watkinsville, GA

Applied N kg ha^{-1}	Applied P kg ha^{-1}	Dry matter Mg ha^{-1} Applied K, kg ha^{-1}			
		0	47	94	188
0	0	4.44	4.77	4.91	4.41
0	25	5.33	5.04	4.73	5.82
0	50	4.91	5.73	4.97	5.51
0	100	4.97	5.49	6.59	5.56
112	0	8.71	8.76	8.20	9.12
112	25	9.74	10.28	10.04	10.19
112	50	9.27	9.77	10.21	9.74
112	100	9.54	10.55	10.82	10.86
224	0	9.86	11.00	11.42	11.72
224	25	11.20	12.63	12.72	12.54
224	50	10.68	12.75	12.99	13.60
224	100	11.58	12.32	13.10	14.43
448	0	9.92	11.89	11.56	13.17
448	25	11.09	14.04	14.52	14.67
448	50	11.69	14.04	14.22	15.05
448	100	12.67	14.00	15.93	16.40

Data adapted from Carreker et al. (1977).

Table 2 Plant N concentration response of coastal bermudagrass to applied N, P, and K at Watkinsville, GA

Applied N kg ha^{-1}	Applied P kg ha^{-1}	Plant N concentration g kg^{-1}			
		Applied K, kg ha^{-1}			
		0	47	94	188
0	0	15.8	15.6	15.7	15.1
0	25	15.5	15.8	15.3	15.5
0	50	15.3	15.7	15.0	15.6
0	100	15.4	15.4	15.8	15.6
112	0	17.8	17.0	17.5	17.3
112	25	17.6	17.3	17.2	16.7
112	50	17.4	16.9	17.0	17.0
112	100	16.9	17.3	17.5	16.9
224	0	21.1	20.1	19.6	19.6
224	25	21.4	19.7	19.7	19.7
224	50	21.8	20.3	19.7	19.7
224	100	20.9	19.8	19.4	19.4
448	0	24.8	24.1	23.8	23.6
448	25	25.3	24.9	24.2	24.5
448	50	26.2	25.1	25.4	21.3
448	100	24.9	24.5	24.6	24.7

Data adapted from Adams et al. (1966).

Table 3 Plant N response of coastal bermudagrass to applied N, P, and K at Watkinsville, GA

Applied N kg ha^{-1}	Applied P kg ha^{-1}	Plant N uptake kg ha^{-1}			
		Applied K, kg ha^{-1}			
		0	47	94	188
0	0	70	74	77	67
0	25	83	80	72	90
0	50	75	90	75	86
0	100	76	85	104	87
112	0	155	149	144	158
112	25	171	178	173	170
112	50	161	165	174	166

continues

Table 3 *continued*

		Plant N uptake kg ha^{-1}			
Applied N kg ha^{-1}	Applied P kg ha^{-1}	Applied K, kg ha^{-1}			
		0	47	94	188
112	100	161	183	189	184
224	0	208	221	225	230
224	25	240	249	252	247
224	50	233	259	255	268
224	100	242	244	254	280
448	0	246	287	275	311
448	25	281	350	351	359
448	50	306	352	361	321
448	100	315	343	392	405

4.26 Adams et al. (1966) and Carreker et al. (1977) reported on response of coastal bermudagrass to applied N, P, and K on Cecil sandy loam at Watkinsville, GA. Data from this study are reported in the tables below. Overman et al. (1991) and Overman and Wilkinson (1995) developed a multiple logistic model for describing response of dry matter and plant nutrient uptake for these data.

a. Plot dry matter yield Y, plant P uptake P_u, and plant P concentration P_c vs. applied phosphorus P for applied potassium of $K = 188$ kg ha^{-1} and applied nitrogen of $N = 0$ and $N = 448$ kg ha^{-1} on linear graph paper.

b. Draw the model curves on (a) from the equations

$$Y = \frac{15.54}{[1 + \exp(0.462 - 0.0122N)][1 + \exp(-1.58 - 0.0410P)]} {[1 + \exp(-1.58 - 0.0212K)]}$$

$$P_u = \frac{37}{[1 + \exp(0.462 - 0.0122N)][1 + \exp(-0.65 - 0.0410P)]} {[1 + \exp(-1.58 - 0.0212K)]}$$

$$P_c = 2.4 \left[\frac{1 + \exp(-1.58 - 0.0410P)}{1 + \exp(-0.65 - 0.0410P)} \right]$$

for the same K and N as in (a).

c. Construct the phase plot of Y and P_c vs. P_u for the same K and N as in (a).

d. Draw the model curves on (c) from the equations

$$K = 188, N = 0 \, \text{kg ha}^{-1}: \quad Y = \frac{9.9 P_u}{9.3 + P_u}$$

$$P_c = 0.94 + 0.101 P_u$$

$$K = 188, N = 448 \, \text{kg ha}^{-1}: \quad Y = \frac{25.4 P_u}{23.9 + K_u}$$

$$P_c = 0.94 + 0.0394 P_u$$

on linear graph paper.

e. Discuss the results. Comment in particular on agreement between the model and data.

Table 1 Yield response of coastal bermudagrass to applied N, P, and K at Watkinsville, GA

Applied N kg ha^{-1}	Applied P kg ha^{-1}	Dry matter Mg ha^{-1} Applied K, kg ha^{-1} 0	47	94	188
0	0	4.44	4.77	4.91	4.41
0	25	5.33	5.04	4.73	5.82
0	50	4.91	5.73	4.97	5.51
0	100	4.97	5.49	6.59	5.56
112	0	8.71	8.76	8.20	9.12
112	25	9.74	10.28	10.04	10.19
112	50	9.27	9.77	10.21	9.74
112	100	9.54	10.55	10.82	10.86
224	0	9.86	11.00	11.42	11.72
224	25	11.20	12.63	12.72	12.54
224	50	10.68	12.75	12.99	13.60
224	100	11.58	12.32	13.10	14.43
448	0	9.92	11.89	11.56	13.17
448	25	11.09	14.04	14.52	14.67
448	50	11.69	14.04	14.22	15.05
448	100	12.67	14.00	15.93	16.40

Data adapted from Carreker et al. (1977).

Table 2 Plant P concentration response of coastal bermudagrass to applied N, P, and K at Watkinsville, GA

Applied N kg ha^{-1}	Applied P kg ha^{-1}	Plant P concentration g kg^{-1} Applied K, kg ha^{-1}			
		0	47	94	188
0	0	2.0	2.0	1.8	1.8
0	25	2.2	2.2	2.2	2.2
0	50	2.3	2.3	2.4	2.2
0	100	2.4	2.5	2.3	2.3
112	0	1.8	1.9	1.7	1.7
112	25	2.2	2.1	2.0	2.0
112	50	2.3	2.2	2.1	2.2
112	100	2.3	2.4	2.3	2.3
224	0	1.8	1.7	1.7	1.7
224	25	2.2	2.1	2.1	2.1
224	50	2.3	2.3	2.1	2.1
224	100	2.3	2.4	2.3	2.3
448	0	1.9	1.7	1.7	1.8
448	25	2.3	2.2	2.0	2.2
448	50	2.3	2.3	2.2	2.2
448	100	2.5	2.4	2.4	2.4

Data adapted from Adams et al. (1966).

Table 3 Plant P response of coastal bermudagrass to applied N, P, and K at Watkinsville, GA

Applied N kg ha^{-1}	Applied P kg ha^{-1}	Plant P uptake kg ha^{-1} Applied K, kg ha^{-1}			
		0	47	94	188
0	0	8.9	9.5	8.8	7.9
0	25	12	11	10	13
0	50	11	13	12	12
0	100	12	14	15	13
112	0	16	17	14	16
112	25	21	22	20	20
112	50	21	21	21	21

Table 3 *continued*

Applied N kg ha^{-1}	Applied P kg ha^{-1}	Plant P uptake kg ha^{-1} Applied K, kg ha^{-1}			
		0	47	94	188
112	100	22	25	25	25
224	0	18	19	19	20
224	25	25	27	27	26
224	50	25	29	27	29
224	100	27	30	30	33
448	0	19	20	20	24
448	25	26	31	29	32
448	50	27	32	31	33
448	100	32	34	38	39

Data adapted from Adams et al. (1966).

4.27 Adams et al. (1966) and Carreker et al. (1977) reported on response of coastal bermudagrass to applied N, P, and K on Cecil sandy loam at Watkinsville, GA. Data from this study are reported in the tables below. Overman et al. (1991) and Overman and Wilkinson (1995) developed a multiple logistic model for describing response of dry matter and plant nutrient uptake for these data.

a. Plot dry matter yield Y, plant K uptake K_u, and plant K concentration K_c vs. applied potassium K for applied phosphorus of $P = 100\,\text{kg ha}^{-1}$ and applied nitrogen of $N = 0$ and $N = 448\,\text{kg ha}^{-1}$ on linear graph paper.

b. Draw the model curves on (a) from the equations

$$Y = \frac{15.54}{[1 + \exp(0.462 - 0.0122N)][1 + \exp(-1.58 - 0.0410P)]} \\ [1 + \exp(-1.58 - 0.0212K)]$$

$$K_u = \frac{270}{[1 + \exp(0.462 - 0.0122N)][1 + \exp(-1.58 - 0.0410P)]} \\ [1 + \exp(0.30 - 0.0212K)]$$

$$K_c = 17.4\left[\frac{1 + \exp(-1.58 - 0.0212K)}{1 + \exp(0.30 - 0.0212K)}\right]$$

for the same P and N as in (a).

c. Construct the phase plot of Y and K_c vs. K_u for the same P and N as in (a).

d. Draw the model curves on (c) from the equations

$$P = 100,\ N = 0\,\mathrm{kg\,ha^{-1}}: \qquad Y = \frac{7.07 K_u}{18.7 + K_u}$$

$$K_c = 2.64 + 0.141 K_u$$

$$P = 100,\ N = 448\,\mathrm{kg\,ha^{-1}}: \qquad Y = \frac{18.1 K_u}{48.1 + K_u}$$

$$K_c = 2.64 + 0.0552 K_u$$

on linear graph paper.

e. Discuss the results. Comment in particular on agreement between the model and data.

f. Discuss adequacy of applied K (particularly at high applied N) as evidenced in Table 3 and in K_c vs. K in the graph of part (b).

Table 1 Yield response of coastal bermudagrass to applied N, P, and K at Watkinsville, GA

Applied N kg ha^{-1}	Applied P kg ha^{-1}	Dry matter Mg ha^{-1} Applied K, kg ha^{-1}			
		0	47	94	188
0	0	4.44	4.77	4.91	4.41
0	25	5.33	5.04	4.73	5.82
0	50	4.91	5.73	4.97	5.51
0	100	4.97	5.49	6.59	5.56
112	0	8.71	8.76	8.20	9.12
112	25	9.74	10.28	10.04	10.19
112	50	9.27	9.77	10.21	9.74
112	100	9.54	10.55	10.82	10.86
224	0	9.86	11.00	11.42	11.72
224	25	11.20	12.63	12.72	12.54
224	50	10.68	12.75	12.99	13.60
224	100	11.58	12.32	13.10	14.43
448	0	9.92	11.89	11.56	13.17
448	25	11.09	14.04	14.52	14.67
448	50	11.69	14.04	14.22	15.05
448	100	12.67	14.00	15.93	16.40

Data adapted from Carreker et al. (1977).

Table 2   Plant K concentration response of coastal bermudagrass to applied N, P, and K at Watkinsville, GA.

Applied N kg ha^{-1}	Applied P kg ha^{-1}	Plant K concentration g kg^{-1} Applied K, kg ha^{-1}			
		0	47	94	188
0	0	11.5	14.2	14.7	15.4
0	25	11.6	13.5	15.6	16.2
0	50	12.0	13.3	14.4	16.1
0	100	12.3	14.9	15.4	15.9
112	0	10.0	12.8	15.7	16.1
112	25	8.6	12.3	14.6	17.4
112	50	9.3	12.4	14.9	17.1
112	100	8.4	13.0	14.7	17.4
224	0	8.8	11.6	15.5	18.0
224	25	8.1	11.5	13.0	17.3
224	50	8.6	12.0	13.3	18.4
224	100	7.4	11.7	13.6	17.1
448	0	8.7	11.4	15.3	19.0
448	25	7.4	11.2	13.4	18.4
448	50	8.4	10.6	17.8	18.0
448	100	8.3	10.8	13.1	17.5

Data adapted from Adams et al. (1966).

Table 3   Plant K response of coastal bermudagrass to applied N, P, and K at Watkinsville, GA

Applied N kg ha^{-1}	Applied P kg ha^{-1}	Plant K uptake kg ha^{-1} Applied K, kg ha^{-1}			
		0	47	94	188
0	0	51	68	72	68
0	25	62	68	74	94
0	50	59	76	72	89
0	100	61	82	101	88
112	0	87	112	129	147
112	25	84	126	147	177
112	50	86	121	152	167

continues

continued

Applied N kg ha^{-1}	Applied P kg ha^{-1}	Plant K uptake kg ha^{-1} Applied K, kg ha^{-1}			
		0	47	94	188
112	100	80	137	159	189
224	0	87	128	177	211
224	25	91	145	165	217
224	50	92	153	173	250
224	100	86	144	178	247
448	0	86	136	177	250
448	25	82	157	195	270
448	50	98	149	253	271
448	100	105	151	209	287

Data adapted from Adams et al. (1966).

4.28 Kamprath (1986) studied the response of corn to applied N at three sites in North Carolina. Data are listed in the table below for grain and total above plant. The three locations and soil types were Clayton, NC [Dothan loamy sand (fine-loamy, siliceous, thermic Plinthic Kandiudults)], Kinston NC [Goldsboro sandy loam (fine-loamy, siliceous, thermic Aquic Paleudults)], and Plymouth NC [Portsmouth very fine sandy loam (fine-loamy over sandy or sandy-skeletal, mixed, thermic Typic Umbraqults)]. The first and second are well-drained soils, while the third is poorly drained. Supplemental irrigation was provided to the Dothan soil only. This exercise will focus on the Dothan soil only.

a. Plot dry matter Y, plant N uptake N_u, and plant N concentration N_c vs. applied N, N, on linear graph paper for the Dothan soil.

b. Draw the model curves on (a) from the equations

$$Y = \frac{A}{1 + \exp(0.30 - 0.0138N)}$$

$$N_u = \frac{A_n}{1 + \exp(0.96 - 0.0138N)}$$

$$N_c = N_{cm}\left[\frac{1 + \exp(0.30 - 0.0138N)}{1 + \exp(0.96 - 0.0138N)}\right]$$

where $A(\text{grain}) = 12.53$ Mg ha^{-1}, $A(\text{total}) = 21.91$ Mg ha^{-1}, $A_n(\text{grain}) = 170$ kg ha^{-1}, $A_n(\text{total}) = 209$ kg ha^{-1}, $N_{cm}(\text{grain}) = 13.6$ g kg^{-1}, and $N_{cm}(\text{total}) = 9.5$ g kg^{-1}.

c. Plot the phase relations Y vs. N_u and N_c vs. N_u on linear graph paper for the Dothan soil.

d. Plot the model curves on (c) from the phase equations

$$Y = \frac{Y_m N_u}{K_n + N_u}$$

$$N_c = \frac{K_n}{Y_m} + \frac{1}{Y_m} N_u$$

where $Y_m(\text{grain}) = 25.9$ Mg ha^{-1}, $K_n(\text{grain}) = 182$ kg ha^{-1}, $Y_m(\text{total}) = 45.4$ Mg ha^{-1}, $K_n(\text{total}) = 224$ kg ha^{-1}.

e. Discuss agreement between the model and data for both grain and total plant.

Note: These data have been analyzed in greater detail by Overman et al. (1994).

Seasonal dry matter yield, plant N removal, and plant N concentration for grain and total plant response to applied N for corn at Clayton, Kinston, and Plymouth, NC

Site (soil)	Part	Applied N kg ha^{-1}	Dry matter Mg ha^{-1}	N removal kg ha^{-1}	N concentration g kg^{-1}
Clayton NC	Grain	0	4.37	47	10.8
(Dothan)		56	7.12	74	10.4
		112	9.77	113	11.6
		168	10.81	135	12.5
		224	11.04	150	13.6
	Total	0	9.09	61	6.7
		56	14.67	92	6.3
		112	18.35	135	7.4
		168	19.11	166	8.7
		224	19.90	188	9.4
Kinston NC	Grain	0	3.00	32	10.7
(Goldsboro)		56	5.47	62	11.3
		112	6.93	87	12.6
		168	7.46	101	13.5
		224	7.57	107	14.1
	Total	0	6.63	36	5.4
		56	10.60	71	6.7
		112	13.10	104	7.9

continues

continued

Site (soil)	Part	Applied N kg ha^{-1}	Dry matter Mg ha^{-1}	N removal kg ha^{-1}	N concentration g kg^{-1}
		168	13.55	120	8.9
		224	13.92	134	9.6
Plymouth NC	Grain	0	4.53	48	10.6
(Portsmouth)		56	6.30	68	10.8
		112	7.65	89	11.6
		168	8.52	104	12.2
		224	9.04	115	12.7
	Total	0	8.93	60	6.7
		56	11.63	83	7.1
		112	13.39	101	7.5
		168	14.68	121	8.2
		224	15.10	133	8.8

Data adapted from Kamprath (1986).

4.29 Given the parametric equations

$$Y_m = \frac{A}{1 - \exp(-\Delta b)}$$

$$K_n = \frac{A_n}{\exp(\Delta b) - 1}$$

derive the linear relationship

$$\frac{Y_m}{A} = 1 + \frac{K_n}{A_n}$$

4.30 Wilson (1995) used the logistic equation to analyze the data of Morrison et al. (1980) for ryegrass at 20 sites in England, as discussed in Exercise 4.22. A summary of some of the results is given in the table below. The analysis was performed with assumption of individual maximum A and intercept b for each site and common response coefficient c for all sites.
   a. Calculate the mean and standard deviation for maximum plant N concentration N_{cm} and differential intercept parameter Δb.
   b. Plot model parameters vs. site on linear graph paper.
   c. Plot the mean ± 2 standard deviations for each parameter on (c).

d. Discuss the variability of the two parameters among sites. Is it reasonable to assume that each parameter is common among sites?

Summary of logistic
model parameters for
ryegrass at 20 sites in
England

Site	N_{cm}, g kg^{-1}	Δb
5	33.9	0.85
6	38.7	1.03
7	39.9	0.59
8	35.9	0.87
9	35.0	0.75
10	34.7	0.76
12	32.9	0.67
13	38.0	0.87
14	35.9	0.85
15	35.9	0.87
16	35.8	0.92
17	33.9	0.75
19	34.9	0.81
20	31.9	0.91
22	35.7	0.72
23	34.7	0.79
25	36.2	0.95
26	34.2	0.86
27	36.4	0.95
28	33.3	0.77

Data from Wilson (1995,
table 4-44).

4.31 Wilson (1995) used the logistic equation to analyze the data of Morrison et al. (1980) for ryegrass at 20 sites in England, as discussed in Exercise 4.22. A summary of some of the results is given in the table below. The analysis was performed with assumption of individual maximum A and intercept b for each site and common response coefficient c for all sites.
   a. Estimate the cumulative frequency distribution for $\Delta b = 0.55, 0.60, 0.65, \ldots, 1.05$; and for $N_{cm} = 32, 33, 34, \ldots, 40$.
   b. Plot normalized frequency distributions F vs. Δb and F vs. N_{cm} on probability paper.

c. Estimate parameters μ and σ from the linearized form

$$z = \mathrm{erf}^{-1}(2F - 1) = -\frac{\mu}{\sqrt{2}\sigma} + \frac{1}{\sqrt{2}\sigma}(\Delta b \text{ or } N_{cm})$$

d. Does the scatter in the values for Δb and N_{cm} appear to be normally distributed?

Summary of logistic
model parameters for
ryegrass at 20 sites in
England

Site	N_{cm}, g kg^{-1}	Δb
5	33.9	0.85
6	38.7	1.03
7	39.9	0.59
8	35.9	0.87
9	35.0	0.75
10	34.7	0.76
12	32.9	0.67
13	38.0	0.87
14	35.9	0.85
15	35.9	0.87
16	35.8	0.92
17	33.9	0.75
19	34.9	0.81
20	31.9	0.91
22	35.7	0.72
23	34.7	0.79
25	36.2	0.95
26	34.2	0.86
27	36.4	0.95
28	33.3	0.77

Data from Wilson (1995,
table 4-44).

4.32 This exercise illustrates the role of parameter σ in the expanded growth
model for perennial grasses harvested on a 6 wk interval. Assume the
parameter values $\mu = 26.0$ wk, $c = 0.2$ wk^{-1}, and $k = 5$.
a. For $\sqrt{2}\sigma = 8.0$ wk, calculate ΔQ_i for calendar times $t =$
0, 6, 12, ..., 42 wk.
b. Calculate $Q_n = \sum_{i=1}^{n} \Delta Q_i$ for each time.

c. Calculate the normalized distribution from $F_n = Q_n/Q_\infty$ for each time, where $Q_\infty = 5.510$.
d. Plot F_n vs. t on probability paper and estimate parameters μ and σ from the line.
e. For $\sqrt{2}\sigma = 16.0\,\text{wk}$, calculate ΔQ_i for calendar times $t = -18, -12, -6, \ldots, 48\,\text{wk}$.
f. Repeat steps (b) through (d) for this case, where $Q_\infty = 26.013$.
g. Is the distribution F_n vs. t linear for these results?
h. Are the values for the parameters μ and σ the same for the output as for the inputs?
i. Discuss the results from this exercise.

REFERENCES

Abramowitz, M., and I. A. Stegun. 1965. *Handbook of Mathematical Functions*. Dover Publications, New York.

Adams, W. E., and M. Stelly. 1962. Fertility requirements of coastal bermudagrass and crimson clover grown on Cecil sandy loam. I. Yield response to fertilization. *J. Range Management* 15:84–87.

Adams, W. E., M. Stelly, R. A. McCreery, H. D. Morris, and C. B. Elkins Jr. 1966. Protein, P, and K composition of coastal bermudagrass and crimson clover. *J. Range Management* 19:301–305.

Alvarez-Sánchez, E., J. D. Etchevers, J. Ortiz, R. Núñez, V. Volke, L. Tijerina, and A. Martínez. 1999. Biomass production and phosphorus accumulation of potato as affected by phosphorus nutrition. *J. Plant Nutr.* 22:205–217.

Baker, J. T., L. H. Allen Jr., and K. J. Boote. 1990. Growth and yield responses of rice to carbon dioxide concentration. *J. Agric. Sci., Camb.* 115:313–320.

Barber, S. A. 1984. *Soil Nutrient Bioavailability*. John Wiley & Sons, New York.

Barrow, J. D. 1994. *Pi in the Sky: Counting, Thinking, and Being*. Little, Brown, and Co., New York.

Beaty, E. R., K. H. Tan, R. A. McCreery, and J. D. Powell. 1980. Yield and N content of closely clipped bahiagrass as affected by N treatments. *Agron. J.* 72:56–60.

Blue, W. G. 1973. Role of pensacola bahiagrass stolon-root systems in fertilizer nitrogen utilization on Leon fine sand. *Agron. J.* 65:88–91.

Bracewell, R. N. 2000. *The Fourier Transform and Its Application*. McGraw-Hill, New York.

Bunch, B. 1989. *Reality's Mirror: Exploring the Mathematics of Symmetry*. John Wiley & Sons, New York.

Burton, G. W., R. N. Gates, and G. J. Gashco. 1997. Response of Pensacola bahiagrass to rates of nitrogen, phosphorus, and potassium fertilizers. *Soil and Crop Sci. Soc. Florida Proc.* 56:31–35.

Burton, G. W., J. E. Jackson, and R. H. Hart. 1963. Effects of cutting frequency and nitrogen on yield, in vitro digestibility, and protein, fiber, and carotene content of coastal bermudagrass. *Agron. J.* 55:500–502.

Carreker, J. R., S. R. Wilkinson, A. P Barnett, and J. E. Box. 1977. Soil and water systems for sloping land. ARS-160. U.S. Government Printing Office, Washington, DC.

Day, J. L., and M. B. Parker. 1985. Fertilizer effects on crop removal of P and K in 'coastal' bermudagrass forage. *Agron. J.* 77:110–114.

Dirac, P. A. M. 1958. *The Principles of Quantum Mechanics*. Oxford University Press, London.

Doss, B. D., D. A. Ashley, O. L. Bennet, and R. M. Patterson. 1966. Interactions of soil moisture, nitrogen, and clipping frequency on yield and nitrogen content of Coastal bermudagrass. *Agron. J.* 58:510–512.

Evans, E. M., L. E. Ensminger, B. D. Doss, and O. L. Bennett. 1961. Nitrogen and moisture requirements of Coastal bermudagrass and Pensacola bahiagrass. Alabama Agric. Exp. Stn. Bulletin 337. Auburn University. Auburn, AL.

Evers, G. W. 1984. Effect of nitrogen fertilizer, clovers, and week control on Coastal bermudagrass and pensacola bahiagrass in southeast Texas. Texas Agric. Exp. Stn. Bulletin MP-1546. Texas A & M University. College Station, TX.

Feynman, R. 1965. *The Character of Physical Law*. MIT Press, Cambridge, MA.

Holt, E. C., and B. E. Conrad. 1986. Influence of harvest frequency and season on bermudagrass cultivar yield and forage quality. *Agron. J.* 78:433–436.

Huneycutt, H. J., C. P. West, and J. M. Phillips. 1988. Responses of bermudagrass, tall fescue and tallfescue-clover to broiler litter and commercial fertilizer. Arkansas Agric. Exp. Stn. Bull. 913. University of Arkansas. Fayetteville, AR.

Icke, V. 1999. *The Force of Symmetry*. Cambridge University Press, Cambridge.

Jean, R. V., and D. Barabé. 1998. *Symmetry in Plants*. World Scientific, Singapore.

Jeffers, R. L. 1955. Response of warm-season permanent-pasture grasses to high levels of nitrogen. *Soil Crop Sci. Soc. Fla. Proc.* 15:231–239.

Johnson, F. H., H. Eyring, and B. J. Stover. 1974. *The Theory of Rate Processes in Biology and Medicine.* John Wiley & Sons, New York.

Kamprath, E. J. 1986. Nitrogen studies with corn on Coastal plain soils. Tech. Bull. 282. North Carolina Agric. Res. Ser. North Carolina State University. Raleigh, NC.

Kamuru, F., S. L. Albrecht, L. H. Allen Jr, and K. T. Shanmugam. 1998. Dry matter and nitrogen accumulation in rice inoculated with a nitrogenase-derepressed mutant of Anabaena variabilis. Agron. J. 90:529-535.

Kimball, B. A. and J. R. Mauney. 1993. Response of cotton to varying CO_2, irrigation, and nitrogen: yield and growth. Agron. J. 85:706-712.

Laidler, K. J. 1965. Chemical Kinetics. McGraw-Hill, New York.

Longair, M. S. 1984. Theoretical Concepts in Physics. Cambridge University Press, Cambridge.

Marschner, H. 1986. Mineral Nutrition of Higher Plants. Academic Press, New York.

McEwen, J., W. Day, I. F. Henderson, A. E. Hutchinson, R. T. Plumb, P. R. Poulton, A. M. Spaull, D. P. Stribley, A. D. Todd, and D. P. Yeoman. 1989. Effects of irrigation, N fertilizer, cutting frequency and pesticides on ryegrass, ryegrass-clover mixtures, clover and lucerne grown on heavy and light land. J. Agric. Sci., Cambridge 112:227-247.

Mengel, K. and E. A. Kirkby. 1987. Principles of Plant Nutrition. International Potash Institute, Bern, Switzerland.

Moore, J. W. and R. G. Pearson. 1981. Kinetics and Mechanism. John Wiley & Sons, New York.

Morrison, J., M. V. Jackson, and P. E. Sparrow. 1980. The response of perennial ryegrass to fertilizer nitrogen in relation to climate and soil. Grassland Research Institute Technical Report No. 27. Berkshire, England.

Mullins, G. L., and C. H. Burmester. 1990. Dry matter, nitrogen, phosphorus, and potassium accumulation by four cotton varieties. *Agron. J.* 82:729-736.

Overman, A. R. 1984. Estimating crop growth rate with land treatment. *J. Env. Engr. Div., Amer. Soc. Civil Engr.* 110:1535-1538.

Overman, A. R. 1995a. Rational basis for the logistic model for forage grasses. *J. Plant Nutr.* 18:995-1012.

Overman, A. R. 1995b. Coupling among applied, soil, root, and top components for forage crop production. *Commun. Soil Sci. Plant Anal.* 26:1179–1202.

Overman, A. R. 1998. An expanded growth model for grasses. *Commun. Soil Sci. Plant Anal.* 29:67–85.

Overman, A. R., E. A. Angley, and S. R. Wilkinson. 1988. A phenomenological model of coastal bermudagrass production. *Agric. Sys.* 29:137–148.

Overman, A. R., E. A. Angley, and S. R. Wilkinson. 1990a. Evaluation of a phenomenological model of coastal bermudagrass production. *Trans. Amer. Soc. Agric. Engr.* 33:443–450.

Overman, A. R., A. Dagan, F. G. Martin, and S. R. Wilkinson. 1991. A nitrogen-phosphorus-potassium model for forage yield of bermudagrass. *Agron. J.* 83:254–258.

Overman, A. R., D. Downey, and S. R. Wilkinson. 1989. Application of simulation models to bahiagrass production. *Commun. Soil Sci. Plant Anal.* 20:1231–1246.

Overman, A. R., and G. W. Evers. 1992. Estimation of yield and nitrogen removal by bermudagrass and bahiagrass. *Trans. Amer. Soc. Agric. Engr.* 35:207–210.

Overman, A. R., F. G. Martin, and S. R. Wilkinson. 1990b. A logistic equation for yield response of forage grass to nitrogen. *Commun. Soil Sci. Plant Anal.* 21:595–609.

Overman, A. R., C. R. Neff, S. R. Wilkinson, and F. G. Martin. 1990c. Water, harvest interval, and applied nitrogen effects on forage yield of bermudagrass and bahiagrass. *Agron. J.* 82:1011–1016.

Overman, A. R., and R. V. Scholtz. 1999. Langmuir-Hinshelwood model of soil phosphorus kinetics. *Commun. Soil Sci. Plant Anal.* 30:109–119.

Overman, A. R., and R. L. Stanley. 1998. Bahiagrass response to applied nitrogen and harvest interval. Commun. *Soil Sci. Plant Anal.* 29:237–244.

Overman, A. R., and S. R. Wilkinson. 1989. Partitioning of dry matter between leaf and stem in coastal bermudagrass. *Agric. Sys.* 30:35–47.

Overman, A. R., and S. R. Wilkinson. 1992. Model evaluation for perennial grasses in the southern United States. *Agron. J.* 84:523–529.

Overman, A. R., and S. R. Wilkinson. 1995. Extended logistic model of forage grass response to applied nitrogen, phosphorus, and potassium. *Trans. Amer. Soc. Agr. Engr.* 38:103–108.

Overman, A. R., S. R. Wilkinson, and G. W. Evers. 1992. Yield response of bermudagrass and bahiagrass to applied nitrogen and overseeded clover. *Agron. J.* 84:998–1001.

Overman, A. R., S. R. Wilkinson, and D. M. Wilson. 1994. An extended model of forage grass response to applied nitrogen. *Agron. J.* 86:617–620.

Overman, A. R., D. M. Wilson, and E. J. Kamprath. 1994. Estimation of yield and nitrogen removal by corn. *Agron. J.* 86:1012–1016.

Pagels, H. 1985. *Perfect Symmetry*. Simon & Schuster, New York.

Prine, G. M., and G. W. Burton. 1956. The effect of nitrogen rate and clipping frequency upon the yield, protein content and certain morphological characteristics of coastal bermudagrass [*Cynodon dactylon* (L.) Pers.]. *Agron. J.* 48:296–301.

Rhoads, F. M., R. L. Stanley Jr., and E. A. Hanlon. 1997. Response of bahiagrass to N, P, and K on an ultisol in north Florida. *Soil Crop Sci. Soc. Fla. Proc.* 56:79–83.

Rojansky, V. 1938. *Introductory Quantum Mechanics*. Prentice-Hall. Englewood Cliffs, NJ.

Rose, M. A. and B. Biernacka. 1999. Seasonal patterns of nutrient and dry weight accumulation in Freeman maple. *HortScience* 34:91–95.

Rothman, M. A. 1985. Conservation laws and symmetry. In: *The Encyclopedia of Physics*. R. M. Besancon (ed.), Van Nostrand Reinhold, New York, pp. 222–226.

Russell, E. J. 1950. *Soil Conditions and Plant Growth*. 8th edition. Longmans, Green & Co., London.

Sartain, J. B. 1992. Phosphorus and zinc influence on bermudagrass growth. *Soil Crop Sci. Soc. Fla. Proc.* 51:39–42.

Vicente-Chandler, J., S. Silva, and J. Figarella. 1959. The effect of nitrogen and frequency of cutting on the yield and composition of three tropical grasses. *Agron. J.* 51:202–206.

Walker, M. E., and D. D. Morey. 1962. Influence of rates of N, P, and K on forage and grain production of Gator rye in South Georgia. Georgia Agric. Exp. Stn. Circular NS 27. University of Georgia. Athens, GA.

Weinberg, S. 1993. *Dreams of a Final Theory*. Random House, New York.

Weyl, H. 1952. *Symmetry*. Princeton University Press, Princeton, N.J.

Wilkinson, S. R., and D. A. Mays. 1979. Mineral nutrition. In: R. C. Buckner and L. P. Bush (eds.), *Tall Fescue*. Monograph No. 20. American Soc. Agron., Madison, WI, pp. 41–73.

Wilson, D. M. 1995. Estimation of dry matter and nitrogen removal by the logistic equation. PhD Dissertation. University of Florida. Gainesville, FL.

Zemanian, A. H. 1987. *Generalized Integral Transformations*. Dover Publications, New York.

5

Pasture Systems

5.1 BACKGROUND

In this chapter we examine the linkage of animal production with forage production, and in turn with applied N. Results from four studies, two with sheep and two with cattle, are examined. Two different mathematical models are discussed.

5.2 QUADRATIC MODEL

The quadratic model is arrived at by looking at data combined with some intuition. For our analysis we define the following variables: $X = $ stocking rate (grazing density), animals per area; $Y = $ animal production (gain), mass per area; and $Y/X = $ specific production, mass per animal. Assume that specific production is related to stocking rate by the linear equation

$$\frac{Y}{X} = a - bX \qquad [5.1]$$

where a and b are model parameters to be determined by regression analysis. Equation [5.1] can be rearranged to the quadratic form

$$Y = aX - bX^2 \qquad [5.2]$$

Now provided that parameters a and b are both positive, Eq. [5.2] will exhibit a maximum given by

$$\frac{dY}{dX} = a - 2bX = 0 \rightarrow X = X_p = \frac{a}{2b} \qquad [5.3]$$

where X_p represents the value of X at peak production, $Y = Y_p$. Substitution of X_p into Eq. [5.2] leads to the value of peak Y,

$$Y_p = aX_p - bX_p^2 = \frac{a^2}{2b} - \frac{a^2}{4b} = \frac{a^2}{4b} \qquad [5.4]$$

Data for the first analysis comes from a study with sheep by Appleton and reported by Owen and Ridgman (1968), as given in Table 5.1. Two levels of applied N were used, labeled simply as low and high. Data are shown in Fig. 5.1, which are certainly consistent with a quadratic model. Linear regression leads to

Low N: $\dfrac{Y}{X} = 27.2 - 0.405X \qquad r = -0.9978 \qquad$ [5.5]

High N: $\dfrac{Y}{X} = 33.1 - 0.382X \qquad r = -0.9986 \qquad$ [5.6]

which are used to draw the lines in Fig. 5.1. It follows that the quadratic equations are given by

Low N: $Y = 27.2X - 0.405X^2 \qquad\qquad\qquad\qquad\qquad$ [5.7]

High N: $Y = 33.1X - 0.382X^2 \qquad\qquad\qquad\qquad\qquad$ [5.8]

which are used to draw the curves in Fig. 5.1. Peak values are estimated as

Low N: $X_p = \dfrac{a}{2b} = 33.6$ sheep ha$^{-1} \rightarrow Y_p = \dfrac{a^2}{4b} = 457$ kg ha^{-1} [5.9]

High N: $X_p = \dfrac{a}{2b} = 43.3$ sheep ha$^{-1} \rightarrow Y_p = \dfrac{a^2}{4b} = 717$ kg ha^{-1} [5.10]

Table 5.1   Sheep gains and stocking rates from Appleton study

N level	X, sheep ha^{-1}	14.8	29.6	44.5	59.3	74.1	88.9
Low	Y/X, kg sheep^{-1}	21.0	16.1	8.4	3.0	−2.5	—
	Y, kg ha^{-1}	310	476	372	178	−181	—
High	Y/X, kg sheep^{-1}	—	22.2	15.4	10.9	4.7	−0.73
	Y, kg ha^{-1}	—	656	685	645	349	−65

Data adapted from Owen and Ridgman (1968).

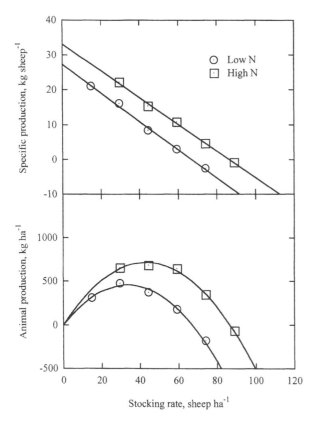

Figure 5.1 Dependence of animal production on stocking rate for sheep from report of Owen and Ridgman (1968). Lines drawn from Eqs. [5.5] and [5.6]; curves drawn from Eqs. [5.7] and [5.8].

The quadratic model appears to describe these data quite well. Peak production is higher for the higher level of applied N, as expected. It is also apparent that too high stocking rate leads to weight loss rather than gain. Both conclusions seem intuitively correct.

A second set of data for sheep comes from a study reported by Owen and Ridgman (1968) as given in Table 5.2 and shown in Fig. 5.2. In this case the line is given by

$$\frac{Y}{X} = 23.4 - 0.256X \qquad r = -0.9757 \qquad\qquad [5.11]$$

Table 5.2 Sheep gains and stocking rates from Hodgson study

X, sheep ha^{-1}	6.2	19.8	33.3	46.9	60.5	74.1
Y/X, kg sheep^{-1}	21.1	18.4	14.2	13.3	9.3	2.4
Y, kg ha^{-1}	131	365	472	624	560	179

Data adapted from Owen and Ridgman (1968).

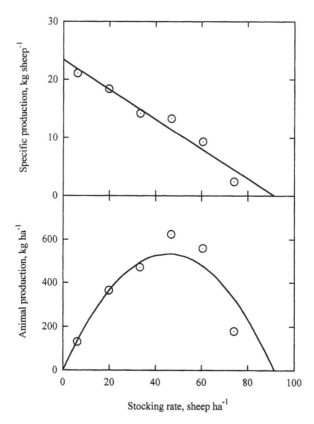

Figure 5.2 Dependence of animal production on stocking rate for sheep from report of Ridgman (1968). Line and curve drawn from Eqs. [5.11] and [5.12], respectively.

with the corresponding curve drawn from

$$Y = 23.4X - 0.256X^2 \tag{5.12}$$

Peak values are estimated from

$$X_p = \frac{a}{2b} = 45.7 \text{ sheep ha}^{-1} \rightarrow Y_p = \frac{a^2}{4b} = 535 \text{ kg ha}^{-1} \tag{5.13}$$

Again the quadratic model describes the general trend in the data.

We next turn attention to a study by Salazar-Diaz (1977) with cattle as given in Table 5.3 and shown in Fig. 5.3. In this study eight stocking rates with steers (*Bos taurus* × *Bos indicus*) were varied with four applied N on pangola digitgrass (*Digitaria decombens* Stent). A weakness of this study was the narrow range of stocking rates, all below the peak rates, which of course leads to uncertainty in parameter estimates. The first step in the analysis is to fit the linear equation to the lowest and highest applied N to obtain

$$N = 168 \text{ kg ha}^{-1}: \quad \frac{Y}{X} = 176 - 8.98X \qquad r = -0.8970 \tag{5.14}$$

$$N = 672 \text{ kg ha}^{-1}: \quad \frac{Y}{X} = 181 - 5.69X \qquad r = -0.9979 \tag{5.15}$$

Since the a parameters are so close together, we choose the value $a = 180$ kg cow^{-1} to use for all applied N levels. Equation [5.1] can now be written as

$$Z \equiv 180 - \frac{Y}{X} = bX \tag{5.16}$$

Table 5.3 Cattle gains and stocking rates from Salazar-Diaz study

N kg ha^{-1}	X, cows ha^{-1}	3.33	4.17	5.00	5.83	6.67	7.50	8.33	9.17
168	Y/X, kg cow^{-1}	142	—	140	—	112	—	—	—
	Y, kg ha^{-1}	474	—	699	—	747	—	—	—
332	Y/X, kg cow^{-1}	—	155	—	140	—	121	—	—
	Y, kg ha^{-1}	—	647	—	817	—	907	—	—
500	Y/X, kg cow^{-1}	—	—	130	—	140	—	134	—
	Y, kg ha^{-1}	—	—	652	—	932	—	1113	—
672	Y/X, kg cow^{-1}	—	—	—	148	—	139	—	129
	Y, kg ha^{-1}	—	—	—	864	—	1043	—	1187

Data adapted from Salazar-Diaz (1977).

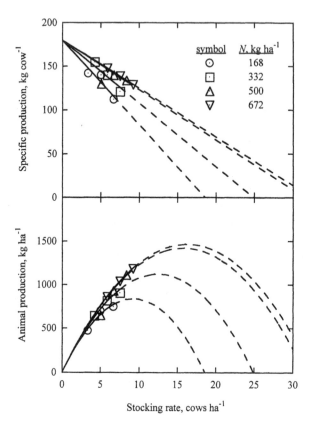

Figure 5.3 Dependence of animal production on stocking rate for cattle from report of Salazar-Diaz (1977). Lines drawn from Eq. [5.18] and curves drawn from Eq. [5.19] with parameter b taken from Table 5.4.

To perform regression analysis so that (X, Z) passes through $(0, 0)$ it is necessary to calculate parameter b for each level from

$$b = \frac{\sum XZ}{\sum X^2} \tag{5.17}$$

Results of this analysis are given in Table 5.4. Lines and curves in Fig. 5.3 are drawn from

$$\frac{Y}{X} = 180 - bX \tag{5.18}$$

$$Y = 180X - bX^2 \tag{5.19}$$

Table 5.4 Model parameters for Salazar-Diaz cattle study

N kg ha^{-1}					b kg cow^{-2}	X_p cows ha^{-1}	Y_p kg ha^{-1}
168	X, cows ha^{-1}	3.33	5.00	6.67	9.68	9.3	837
	Z, kg cow^{-1}	38	40	68			
332	X, cows ha^{-1}	4.17	5.83	7.50	7.22	12.5	1120
	Z, kg cow^{-1}	38	40	68			
500	X, cows ha^{-1}	—	5.00	6.67	5.70	15.8	1420
	Z, kg cow^{-1}	—	40	46			
672	X, cows ha^{-1}	5.83	5.00	6.67	5.53	16.3	1470
	Z, kg cow^{-1}	32	41	51			

where values for model parameter b are taken from Table 5.4 for each applied N. Dependence of X_p on applied N is shown in Fig. 5.4. This response is reminiscent of the logistic equation, so it seems reasonable to assume this form

$$X_p = \frac{A}{1 + \exp(b - cN)} \qquad [5.20]$$

where parameters A, b, and c are estimated by regression analysis. By visual inspection we choose $A = 17.0$ cows ha^{-1}. Equation [5.20] can then be rearranged to the linear form

$$\ln\left(\frac{17.0}{X_p} - 1\right) = b - cN = 0.88 - 0.0062N \qquad r = -0.9838 \qquad [5.21]$$

as shown in Fig. 5.5 The curve in Fig. 5.4 is drawn from

$$X_p = \frac{17.0}{1 + \exp(0.88 - 0.0062N)} \qquad [5.22]$$

A more definitive study with Aberdeen Angus cattle (*Bos taurus*) with kikuyugrass (*Pennisetum clandestinum*) was reported by Mears and Humphreys (1974), as given in Table 5.5 and shown in Fig. 5.6. Since it appears that Y/X vs. X converges to a common a value, we choose parameter $a = 220$ kg cow^{-1}. Following the procedure of the previous case, it is convenient to define

$$Z \equiv 220 - \frac{Y}{X} = bX \qquad [5.23]$$

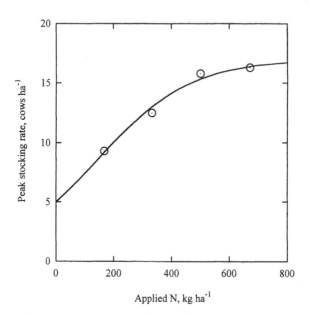

Figure 5.4 Dependence of peak stocking rate on applied N for cattle from report of Salazar-Diaz (1977). Curve drawn from Eq. [5.22].

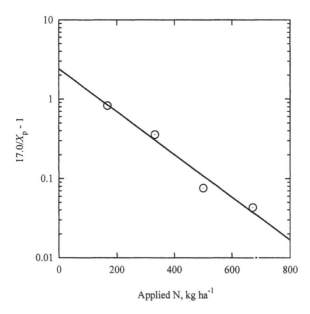

Figure 5.5 Semilog plot of peak stocking rate vs. applied N for cattle from report of Salazar-Diaz (1977). Line drawn from Eq. [5.21].

Table 5.5 Cattle gains and stocking rates from Mears and Humphreys study

N kg ha^{-1}	X, cows ha^{-1}	2.2	3.3	5.0	7.5	11.1	16.6
0	Y/X, kg cow^{-1}	130	116	53	—	—	—
	Y, kg ha^{-1}	286	383	265	—	—	—
134	Y/X, kg cow^{-1}	—	136	101	43	—	—
	Y, kg ha^{-1}	—	449	505	323	—	—
336	Y/X, kg cow^{-1}	—	—	165	118	64	—
	Y, kg ha^{-1}	—	—	825	885	710	—
672	Y/X, kg cow^{-1}	—	—	—	141	95	37
	Y, kg ha^{-1}	—	—	—	1056	1055	614

Data adapted from Mears and Humphreys (1974).

Parameter b is again calculated by Eq. [5.17], as given in Table 5.6. Lines and curves in Fig. 5.6 are drawn from

$$\frac{Y}{X} = 220 - bX \tag{5.24}$$

$$Y = 220X - bX^2 \tag{5.25}$$

where values for parameter b are taken from Table 5.6 for each applied N. Dependence of X_p on applied N is shown in Fig. 5.7, which exhibits logistic response. By visual inspection $A = 10.5$ cows ha^{-1}, which leads to the linearized equation

$$\ln\left(\frac{10.5}{X_p} - 1\right) = b - cN = 0.87 - 0.0058N \qquad r = -0.9973 \tag{5.26}$$

as shown in Fig. 5.8. The curve in Fig. 5.7 is drawn from

$$X_p = \frac{10.5}{1 + \exp(0.87 - 0.0058N)} \tag{5.27}$$

The similarity of logistic parameters b and c in Eqs. [5.22] and [5.27] should be noted. This fact lends support to the linkage between animal production, forage production, and applied N through the quadratic and logistic models.

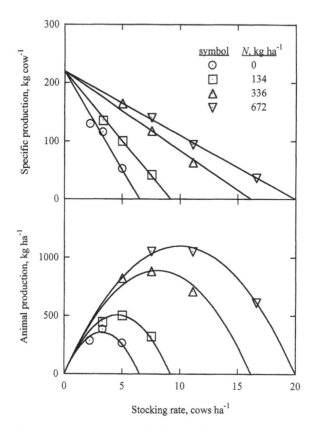

Figure 5.6 Dependence of animal production on stocking rate for cattle from report of Humphreys (1972). Lines drawn from Eq. [5.24] and curves drawn from Eq. [5.25] with parameter b taken from Table 5.6.

5.3 LINEAR EXPONENTIAL MODEL

While the quadratic model appears to describe data for sheep and cattle production quite well, we now discuss a second model which may have merit for describing cattle production. We choose to call it the linear exponential model, a name which will become apparent shortly.

We start by assuming that specific production Y/X is related to stocking rate X by the Gaussian function

$$\frac{Y}{X} = A \exp(-bX^2) \qquad\qquad [5.28]$$

Table 5.6 Quadratic model parameters for Mears and Humphreys cattle study

N kg ha^{-1}					b kg cow^{-2}	X_p cows ha^{-1}	Y_p kg ha^{-1}
0	X, cows ha^{-1}	2.2	3.3	5.0	33.8	3.25	358
	Z, kg cow^{-1}	90	104	167			
134	X, cows ha^{-1}	3.3	5.0	7.5	23.9	4.60	506
	Z, kg cow^{-1}	84	119	177			
336	X, cows ha^{-1}	5.0	7.5	11.1	13.6	8.09	890
	Z, kg cow^{-1}	55	102	156			
672	X, cows ha^{-1}	7.5	11.1	16.6	11.0	10.0	1100
	Z, kg cow^{-1}	79	125	183			

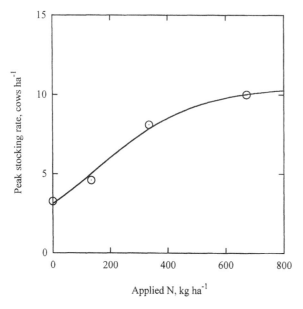

Figure 5.7 Dependence of peak stocking rate on applied N for cattle from report of Humphreys (1972). Curve drawn from Eq. [5.27].

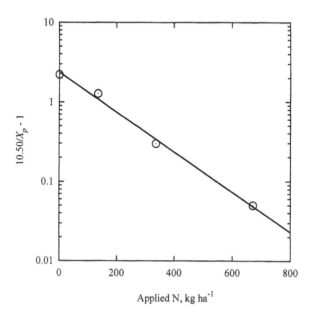

Figure 5.8   Semilog plot of peak stocking rate vs. applied N for cattle from report of Humphreys (1972). Line drawn from Eq. [5.26].

where A and b are model parameters. Equation [5.28] can be rearranged to the form

$$Y = AX \exp(-bX^2) \tag{5.29}$$

For the case where parameters A and b are positive, the point of maximum production, $X = \sigma$, is given by

$$\frac{dY}{dX} = A(1 - 2bX^2)\exp(-bX^2) = 0 \;\rightarrow\; b = \frac{1}{2\sigma^2} \tag{5.30}$$

The symbol σ is used because this happens to be the point of inflection of the Gaussian distribution, which is also the standard deviation of the distribution. Equation [5.29] can now be written as

$$Y = A\sigma\left(\frac{X}{\sigma}\right)\exp\left(-\frac{X^2}{2\sigma^2}\right) \tag{5.31}$$

Introduction of the dimensionless variables

$$\xi = \frac{X}{\sigma} \quad\text{and}\quad \phi = \frac{Y}{A\sigma} \tag{5.32}$$

into Eq. [5.31] leads to

$$\phi = \xi \exp\left(-\frac{\xi^2}{2}\right) \tag{5.33}$$

Equation [5.33] represents the Hermite function, which has played prominent role in quantum mechanics (Pauling and Wilson, 1963; Rojansky, 1938). The extrema (points of maximum or minimum) for Eq. [5.33] are given by

$$\frac{d\phi}{d\xi} = (1 - \xi^2) \exp\left(-\frac{\xi^2}{2}\right) = 0 \rightarrow \xi = -1, +1 \tag{5.34}$$

where -1 and $+1$ represent minimum and maximum points of ϕ, respectively. As a point of interest, we note that

$$\frac{d^2\phi}{d\xi^2} = \frac{d}{d\xi}\left[(1 - \xi^2)\exp\left(-\frac{\xi^2}{2}\right)\right] = -(3 - \xi^2)\phi \tag{5.35}$$

which upon rearrangement becomes

$$\frac{d^2\phi}{d\xi^2} + (3 - \xi^2)\phi = 0 \tag{5.36}$$

Equation [5.36] is the Schrödinger equation for a linear harmonic oscillator with an eigenvalue of 3. The solution to Eq. [5.36] is

$$\phi = AH(\xi) \exp\left(-\frac{\xi^2}{2}\right) \tag{5.37}$$

where the Hermite function H(ξ) satisfies Hermite's differential equation (Riley, 1997, p. 433)

$$\frac{d^2H}{d\xi^2} - 2\xi\frac{dH}{d\xi} + 2H = 0 \tag{5.38}$$

where A is an arbitrary constant. The first order Hermite function is given by (Pauling and Wilson, 1963, p. 81)

$$H_1(\xi) = 2\xi \tag{5.39}$$

so that

$$\phi = \frac{1}{2}H(\xi) \exp\left(-\frac{\xi^2}{2}\right) = \xi \exp\left(-\frac{\xi^2}{2}\right) \tag{5.40}$$

The value of A in Eq. [5.37] is chosen so as to normalize the function

$$\int_0^\infty \xi \exp\left(-\frac{\xi^2}{2}\right) d\xi = -\exp\left(-\frac{\xi^2}{2}\right)\Big|_0^\infty = 1 \qquad [5.41]$$

Equation [5.28] can be linearized to the form of $\ln(Y/X)$ vs. X^2

$$\ln\left(\frac{Y}{X}\right) = \ln A - bX^2 \qquad [5.42]$$

to obtain the following equations for data in Table 5.5:

$N = 0$: $\qquad\qquad \ln\left(\frac{Y}{X}\right) = 5.16 - 0.0465X^2 \qquad r = -0.9838 \quad [5.43]$

$N = 134\,\text{kg ha}^{-1}$: $\quad \ln\left(\frac{Y}{X}\right) = 5.21 - 0.0257X^2 \qquad r = -0.9984 \quad [5.44]$

$N = 336\,\text{kg ha}^{-1}$: $\quad \ln\left(\frac{Y}{X}\right) = 5.33 - 0.00956X^2 \qquad r = -0.9992 \quad [5.45]$

$N = 672\,\text{kg ha}^{-1}$: $\quad \ln\left(\frac{Y}{X}\right) = 5.29 - 0.00611X^2 \qquad r = -0.99994 \quad [5.46]$

Since the average value of $\ln A = 5.25$, we fix $A = 190\,\text{kg cow}^{-1}$ for all applied N. It is convenient to define the dimensionless variable F by

$$F \equiv -\ln\left[\left(\frac{1}{A}\right)\left(\frac{Y}{X}\right)\right] = +bX^2 \qquad [5.47]$$

Then parameter b can be estimated from

$$b = \frac{\sum FX^2}{\sum X^4} \qquad [5.48]$$

and parameter σ follows from

$$\sigma = \left(\frac{1}{2b}\right)^{1/2} \qquad [5.49]$$

Results of this analysis are shown in Table 5.7. The curves in Fig. 5.9 are drawn from

$$\frac{Y}{X} = 190 \exp\left(-\frac{X^2}{2\sigma^2}\right) \qquad [5.50]$$

$$Y = 190X \exp\left(-\frac{X^2}{2\sigma^2}\right) \qquad [5.51]$$

Table 5.7 Linear exponential model parameters for Mears and Humphreys cattle study

N kg ha⁻¹	X, cows ha⁻¹ X^2, cows² ha⁻²	2.2 4.84	3.3 10.9	5.0 25.0	7.5 56.2	11.1 123.2	16.6 275.6	σ cows ha⁻¹
0	Y/X, kg cow⁻¹	130	116	53	—	—	—	
	F	0.379	0.494	1.277	—	—	—	3.13
134	Y/X, kg cow⁻¹	—	136	101	43	—	—	
	F	—	0.334	0.631	1.487	—	—	4.35
336	Y/X, kg cow⁻¹	—	—	165	118	64	—	
	F	—	—	0.142	0.476	1.088	—	7.60
672	Y/X, kg cow⁻¹	—	—	—	141	95	37	
	F	—	—	—	0.298	0.693	1.635	9.23

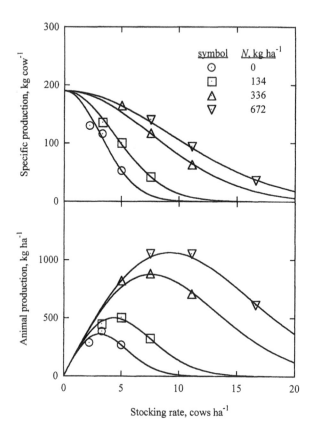

Figure 5.9 Dependence of animal production on stocking rate for cattle from report of Humphreys (1972). Curves drawn from Eqs. [5.50] and [5.51] with parameter b taken from Table 5.7.

with values of σ for each applied N from Table 5.7. The corresponding semilog plot of Y/X vs. X^2 is shown in Fig. 5.10, which lends further support for the linear exponential model. Dependence of peak stocking rate σ on applied N is shown in Fig. 5.11, where the curve is drawn from the logistic equation

$$\sigma = \frac{9.50}{1 + \exp(0.84 - 0.00649N)} \qquad [5.52]$$

The linearized form of Eq. [5.52]

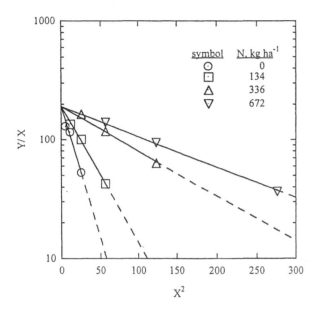

Figure 5.10   Semilog plot of specific production vs. square of stocking rate for cattle from report of Humphreys (1972). Lines drawn from Eq. [5.51] with parameter b taken from Table 5.7.

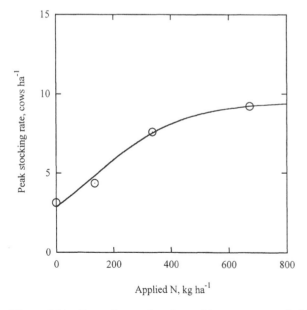

Figure 5.11   Dependence of peak stocking rate on applied N for cattle from report of Humphreys (1972). Curve drawn from Eq. [5.52].

$$\ln\left(\frac{9.50}{\sigma} - 1\right) = b - cN = 0.84 - 0.00649N \qquad r = -0.9974 \qquad [5.53]$$

is used to draw the line in the semilog plot Fig. 5.12. As with the quadratic model, the logistic equation provides excellent correlation of peak stocking rate with applied N for the linear exponential model. Furthermore, the logistic parameters are very similar for the two production models.

5.4 SUMMARY

We now attempt to provide a rational basis for the two grazing models. First for the quadratic model. The intercept parameter a in Eq. [5.1] represents maximum specific animal production near zero stocking rate, which is given mathematically by

$$\lim_{X\to 0}\left(\frac{Y}{X}\right) = a \qquad\qquad [5.54]$$

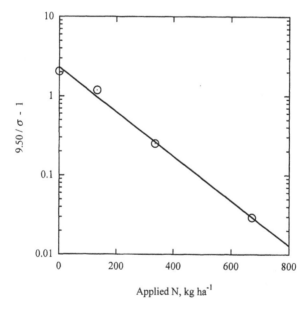

Figure 5.12 Semilog plot of peak stocking rate vs. applied N for cattle from report of Humphreys (1972). Line drawn from Eq. [5.53]

Rate of loss of specific production with increased stocking rate is given by

$$\frac{d}{dX}\left(\frac{Y}{X}\right) = -b = \text{constant} \qquad [5.55]$$

Parameter b can be viewed as the coefficient of competition. The linear exponential model presents a different picture. In this case the intercept parameter A in Eq. [5.28] also represents maximum specific animal production near zero stocking rate, as given by

$$\lim_{X \to 0}\left(\frac{Y}{X}\right) = A \qquad [5.56]$$

Now rate of loss of specific production with increased stocking rate is given by

$$\frac{d}{dX}\left(\frac{Y}{X}\right) = -2AbX \exp(-bX^2) = -2bX\left(\frac{Y}{X}\right) \qquad [5.57]$$

which depends on stocking rate X and specific production Y/X. Again, parameter b can be viewed as the coefficient of competition.

While there are some similarities between the two models, there are distinct differences. The quadratic model allows for negative production (weight loss) at sufficiently high stocking rates, whereas the linear exponential model always predicts positive production. Data from the sheep studies show this loss characteristic (Figs. 5.1 and 5.2). The linear exponential model exhibits a "shoulder" at low stocking rates (Fig. 5.6), which means that the competition factor is small at low stocking rates. Both models provide excellent correlation of peak stocking rate with applied N through the logistic equation. So, both models are presented for further evaluation by other scientists. Discussion of models for grazing systems can be found in Jones and Sandland (1974), including the quadratic model of Section 5.2. Since there has been disagreement about the form of the response function at very low stocking rates, two different models have been presented.

EXERCISES

5.1 Nguyen and Goh (1992) measured yields and plant P removal by ryegrass under grazing by sheep at Canterbury, New Zealand. Measurements were taken in both camp and non-camp areas. Data are given below.

Phosphorus recovery by perennial ryegrass from
grazed areas at Canterbury, New Zealand

Type	P kg ha^{-1}	Y Mg ha^{-1}	P_u kg ha^{-1}	P_c g kg^{-1}
Camp	0	5.09	12.5	2.46
	17.5	13.10	48.5	3.70
	35.0	15.54	70.4	4.53
Noncamp	0	3.25	7.61	2.34
	17.5	9.27	32.8	3.54
	35.0	10.53	48.3	4.59

Data adapted from Nguyen and Goh (1992).

a. Plot dry matter Y, plant P uptake (P_u), and plant P concentration (P_c) vs. applied P for the camp area on linear graph paper.

b. Draw the response curves on (a) from the logistic equations

$$Y = \frac{16.00}{1 + \exp(0.70 - 0.125P)}$$

$$P_u = \frac{75.0}{1 + \exp(1.60 - 0.125P)}$$

$$P_c = 4.69 \left[\frac{1 + \exp(0.70 - 0.125P)}{1 + \exp(1.60 - 0.125P)} \right]$$

c. Show the phase plot for Y vs. P_u and P_c vs. P_u for the camp area on linear graph paper.

d. Calculate and plot the phase relations on (c).

e. Discuss the results. Compare the maximum potential dry matter yield to maximum measured yield.

5.2 Nguyen and Goh (1992) measured dry matter and plant P accumulation by ryegrass under grazing by sheep at Canterbury, New Zealand. Measurements were taken in both camp and noncamp areas. Data are given below for the camp area with no applied P. Time is referenced to Jan. 1.

Dry matter and plant P accumulation (camp area, no applied P) with time for perennial ryegrass at Canterbury, New Zealand

t wk	ΔY Mg ha^{-1}	Y Mg ha^{-1}	F_y	ΔP_u kg ha^{-1}	P_u kg ha^{-1}	F_p	F_{avg}	z
−17.3		0	0		0	0	0	—
	0.15			0.31				
−13.0		0.15	0.029		0.31	0.025	0.027	−1.36
	0.60			1.50				
−8.4		0.75	0.147		1.81	0.145	0.146	−0.745
	0.78			1.87				
−4.3		1.53	0.300		3.68	0.294	0.297	−0.376
	0.97			2.63				
0		2.50	0.491		6.31	0.504	0.498	0.00
	1.02			2.44				
4.4		3.52	0.692		8.75	0.699	0.696	0.363
	0.76			2.14				
13.0		4.28	0.841		10.89	0.870	0.855	0.748
	0.63			1.25				
17.3		4.91	0.964		12.14	0.968	0.966	1.29
	0.18			0.38				
21.7		5.09	1		12.52	1	1	—

Data adapted from Nguyen and Goh (1992).

a. Plot F vs. time for dry matter and plant P on probability paper.
b. Estimate parameters μ and σ from the graph.
c. Calculate cumulative dry matter and plant P uptake from

$$Y = \frac{5.09}{2}\left(1 + \operatorname{erf}\frac{t - \mu}{\sqrt{2}\sigma}\right) \qquad P_u = \frac{12.52}{2}\left(1 + \operatorname{erf}\frac{t - \mu}{\sqrt{2}\sigma}\right)$$

d. Plot (c) on linear graph paper.
e. Plot F_p vs. F_y on linear graph paper.
f. Discuss the results. Do (a) and (e) follow straight lines?

REFERENCES

Jones, R. J., and R. L. Sandland. 1974. The relation between animal gain and stocking rate: Derivation of the relation from the results of grazing trials. *J. Agric. Sci., Camb.* 83:335–342.

Mears, P. T., and L. R. Humphreys. 1974. Nitrogen response and stocking rate of *Pennisetum clandestinum* pastures. II. Cattle growth. *J. Agric. Sci., Camb.* 83:469–478.

Nguyen, M. L., and K. M. Goh. 1992. Nutrient cycling and losses based on a mass-balance model in grazed pastures receiving long-term superphosphate applications in New Zealand. I. Phosphorus. *J. Agric. Sci., Camb.* 119:89–106.

Owen, J. B., and W. J. Bridgman. 1968. The design and interpretation of experiments to study animal production from grazed pastures. *J. Agric. Sci., Camb.* 71:327–335.

Pauling, L., and E. B. Wilson. 1963. *Introduction to Quantum Mechanics with Applications to Chemistry.* Dover Publications, New York.

Riley, K. F., M. P. Hobson, and S. J. Bence. 1997. *Mathematical Methods for Physics and Engineering.* Cambridge Univ. Press, Cambridge, England.

Rojansky, V. 1938. *Introductory Quantum Mechanics.* Prentice-Hall, Englewood Cliffs, NJ.

Salazar-Diaz, J. M. 1977. Effects of nitrogen fertilization and stocking rate on forage and beef production from pangola digitgrass (*Digitaria decumbens* Stent) pastures in Colombia. PhD Dissertation. University of Florida, Gainesville, FL.

6

Nonlinear Regression for Mathematical Models

6.1 BACKGROUND

This is perhaps the most tedious chapter of the text. Some readers will be content to skip this material and use the models anyway. Others will find useful insight into the mathematics of modeling in this chapter. Either approach has its place, and the reader should not feel guilty at ignoring this topic.

There are many approaches to numerical procedures for regression analysis. Some references which the authors have found useful include Adby and Dempster (1974), Bard (1974), Bates and Watts (1988), Draper and Smith (1981), Freund and Littell (1991), Hosmer and Lemeshow (1989), Kleinbaum (1994), Ratkowsky (1983), and Seber and Wild (1989). The second-order Newton-Raphson method (Adby and Dempster, 1974) of nonlinear regression has proven particularly useful.

In the context of regression analysis, the term *nonlinear* refers to parameters in the model, and in which one or more of the parameters occur in nonlinear form. This is made clear in the next section. We normally think of nonlinear in the context of variables in a model. The distinction between these two should be clear in the reader's mind.

6.2 LOGISTIC MODEL

The simple logistic model is given by

$$Y = \frac{A}{1 + \exp(b - cN)} \tag{6.1}$$

where Y = seasonal dry matter yield, Mg ha^{-1}; N = applied nitrogen, kg ha^{-1}; A = maximum seasonal dry matter yield, Mg ha^{-1}; b = intercept parameter for dry matter; c = N response coefficient for dry matter, ha kg^{-1}. For regression purposes, Eq. [6.1] is linear in parameter A, but nonlinear in parameters b and c. The challenge is to optimize the fit of Eq. [6.1] to a given set of data, which means optimizing the choice of parameters A, b, and c. The standard criteria is to minimize the error of the sum of squares of deviations between the data and the model (Draper and Smith, 1981).

Error sum of squares E between the data and the model is defined by

$$E = \sum [Y - \hat{Y}]^2 = \sum \left[Y - \frac{A}{1 + \exp(b - cN)} \right]^2 \tag{6.2}$$

where Y = measured yield and $\hat{Y}$ = estimated yield (Eq. [6.1]). The challenge is to choose parameters A, b, and c to minimize E for a particular set of data. A necessary condition for minimum E is that simultaneously

$$\frac{\partial E}{\partial A} = 0, \quad \frac{\partial E}{\partial b} = 0, \quad \frac{\partial E}{\partial c} = 0 \tag{6.3}$$

The partial derivative w.r.t. (with respect to) A is given by

$$\frac{\partial E}{\partial A} = 2 \sum \left[Y - \frac{A}{1 + \exp(b - cN)} \right] \left[\frac{-1}{1 + \exp(b - cN)} \right]$$

$$= -2 \left\{ \sum \frac{Y}{1 + \exp(b - cN)} - A \sum \frac{1}{[1 + \exp(b - cN)]^2} \right\} = 0 \tag{6.4}$$

which leads to

$$A = \frac{\sum \dfrac{Y}{1 + \exp(b - cN)}}{\sum \dfrac{1}{[1 + \exp(b - cN)]^2}} \tag{6.5}$$

According to Eq. [6.5] optimum A can be calculated provided values of b and c are assumed. The partial derivative w.r.t. b is given by

$$\frac{\partial E}{\partial b} = 2 \sum \left\{ Y - \frac{A}{1 + \exp(b - cN)} \right\} \left\{ \frac{A \exp(b - cN)}{[1 + \exp(b - cN)]^2} \right\}$$

$$= 2A \left\{ \sum \frac{Y \exp(b - cN)}{[1 + \exp(b - cN)]^2} - A \sum \frac{\exp(b - cN)}{[1 + \exp(b - cN)]^3} \right\} \qquad [6.6]$$

Since Eq. [6.6] leads to an implicit function in b, an iterative technique is required to obtain the optimum value of b. We choose the second-order Newton-Raphson method (Adby and Dempster, 1974), which requires values for second-derivatives. The second-order Newton-Raphson method uses a truncated Taylor series such that

$$\left(\frac{\partial E}{\partial b}\right)_{b+\Delta b, c+\Delta c} = \left(\frac{\partial E}{\partial b}\right)_{b,c} + \left(\frac{\partial^2 E}{\partial b^2}\right)_{b,c} \Delta b + \left(\frac{\partial^2 E}{\partial b \, \partial c}\right)_{b,c} \Delta c + \cdots$$

$$\approx \left(\frac{\partial E}{\partial b}\right)_{b,c} + \left(\frac{\partial^2 E}{\partial b^2}\right)_{b,c} \Delta b + \left(\frac{\partial^2 E}{\partial b \, \partial c}\right)_{b,c} \Delta c = 0 \qquad [6.7]$$

$$\left(\frac{\partial E}{\partial c}\right)_{b+\Delta b, c+\Delta c} = \left(\frac{\partial E}{\partial c}\right)_{b,c} + \left(\frac{\partial^2 E}{\partial c \, \partial b}\right)_{b,c} \Delta b + \left(\frac{\partial^2 E}{\partial c^2}\right)_{b,c} \Delta c + \cdots$$

$$\approx \left(\frac{\partial E}{\partial c}\right)_{b,c} + \left(\frac{\partial^2 E}{\partial c \, \partial b}\right)_{b,c} \Delta b + \left(\frac{\partial^2 E}{\partial c^2}\right)_{b,c} \Delta c = 0 \qquad [6.8]$$

Equations [6.7] and [6.8] can be reduced to the two simultaneous equations

$$\left(\frac{\partial^2 E}{\partial b^2}\right)_{b,c} \Delta b + \left(\frac{\partial^2 E}{\partial b \, \partial c}\right)_{b,c} \Delta c = -\left(\frac{\partial E}{\partial b}\right)_{b,c} \qquad [6.9]$$

$$\left(\frac{\partial^2 E}{\partial c \, \partial b}\right)_{b,c} \Delta b + \left(\frac{\partial^2 E}{\partial c^2}\right)_{b,c} \Delta c = -\left(\frac{\partial E}{\partial c}\right)_{b,c} \qquad [6.10]$$

This system of equations takes the matrix form

$$\begin{bmatrix} H_{bb} & H_{bc} \\ H_{cb} & H_{cc} \end{bmatrix} \begin{bmatrix} \Delta b \\ \Delta c \end{bmatrix} = \begin{bmatrix} -J_b \\ -J_c \end{bmatrix} \qquad [6.11]$$

where elements of the Hessian matrix $[H]$ are given by

$$H_{bb} = \frac{\partial^2 E}{\partial b^2}, \quad H_{bc} = \frac{\partial^2 E}{\partial b \partial c} = \frac{\partial^2 E}{\partial c \partial b} = H_{cb}, \quad H_{cc} = \frac{\partial^2 E}{\partial c^2} \qquad [6.12]$$

and elements of the Jacobian vector $[J]$ are given by

$$J_b = \frac{\partial E}{\partial b}, \quad J_c = \frac{\partial E}{\partial c} \tag{6.13}$$

Equation [6.11] can be solved by Cramer's rule to obtain

$$\Delta b = \frac{D_b}{D} \tag{6.14}$$

$$\Delta c = \frac{D_c}{D} \tag{6.15}$$

where the determinants in Eqs. [6.14] and [6.15] are given by

$$D = \begin{vmatrix} H_{bb} & H_{bc} \\ H_{cb} & H_{cc} \end{vmatrix} = H_{bb}H_{cc} - H_{bc}^2 \tag{6.16}$$

$$D_b = \begin{vmatrix} -J_b & H_{bc} \\ -J_c & H_{cc} \end{vmatrix} = -J_b H_{cc} + J_c H_{bc} \tag{6.17}$$

$$D_c = \begin{vmatrix} H_{bb} & -J_b \\ H_{cb} & -J_c \end{vmatrix} = -J_c H_{bb} + J_b H_{cb} \tag{6.18}$$

Convergence of the Newton-Raphson procedure requires that $D > 0$ (positive definite), which constitutes the sufficient condition for a minimum. In this case the procedure is self-correcting.

The second derivative w.r.t. b is given by

$$\frac{\partial^2 E}{\partial b^2} = 2A \left\{ \sum \frac{Y \exp(b - cN)}{[1 + \exp(b - cN)]^2} - 2 \sum \frac{Y \exp 2(b - cN)}{[1 + \exp(b - cN)]^3} \right. \\ \left. - A \sum \frac{\exp(b - cN)}{[1 + \exp(b - cN)]^3} + 3A \sum \frac{\exp 2(b - cN)}{[1 + \exp(b - cN)]^4} \right\} \tag{6.19}$$

while the partial derivatives w.r.t. c are given by

$$\frac{\partial E}{\partial c} = 2 \sum \left\{ Y - \frac{A}{[1 + \exp(b - cN)]} \right\} \left\{ \frac{AN \exp(b - cN)}{[1 + \exp(b - cN)]^2} \right\} \\ = -2 \left\{ \sum \frac{YN \exp(b - cN)}{[1 + \exp(b - cN)]^2} - A \sum \frac{N \exp(b - cN)}{[1 + \exp(b - cN)]^3} \right\} \tag{6.20}$$

and

$$\frac{\partial^2 E}{\partial c^2} = 2A \left\{ \begin{array}{l} \sum \dfrac{YN^2 \exp(b-cN)}{[1+\exp(b-cN)]^2} - 2\sum \dfrac{YN^2 \exp 2(b-cN)}{[1+\exp(b-cN)]^3} \\[3mm] -A\sum \dfrac{N^2 \exp(b-cN)}{[1+\exp(b-cN)]^3} + 3A\sum \dfrac{N^2 \exp 2(b-cN)}{[1+\exp(b-cN)]^4} \end{array} \right\}$$

[6.21]

The cross derivatives are given by

$$\frac{\partial^2 E}{\partial b \partial c} = \frac{\partial^2 E}{\partial c \partial b}$$

$$= -2A \left\{ \begin{array}{l} \sum \dfrac{YN \exp(b-cN)}{[1+\exp(b-cN)]^2} - 2\sum \dfrac{YN \exp 2(b-cN)}{[1+\exp(b-cN)]^3} \\[3mm] -A\sum \dfrac{N \exp(b-cN)}{[1+\exp(b-cN)]^3} + 3A\sum \dfrac{N \exp 2(b-cN)}{[1+\exp(b-cN)]^4} \end{array} \right\}$$

[6.22]

Corrections for estimates of b and c follow from

$$b \text{ (new)} = b \text{ (old)} + \Delta b \qquad\qquad [6.23]$$

$$c \text{ (new)} = c \text{ (old)} + \Delta c \qquad\qquad [6.24]$$

Iteration of the procedure continues until

$$\left| \frac{\Delta b}{b} \right| \le \varepsilon, \quad \left| \frac{\Delta c}{c} \right| \le \varepsilon \qquad\qquad [6.25]$$

where ε is an arbitrary number, typically 10^{-3}–10^{-5}. Again, convergence of the procedure requires that the determinant of the coefficient matrix be positive definite.

Initial estimates of the nonlinear parameters b and c are required. These can be obtained either graphically or by linear regression of the linearized form of the model.

There are numerous cases where comparisons between or among data sets are desirable. In such cases, some parameters may be common to all sets while others may differ from one set to another. One example would be comparison of yields between a "wet" year and a "dry" year. In this case the A parameter may differ between years, while b and c may be common to both years. For this example individual A values would be obtained by summing Eq. [6.6] for each set of values. Parameters b and c would be estimated by summing over all of the data. Another example would be analysis of data for yield and plant N uptake with

$$N_u = \frac{A_n}{1 + \exp(b_n - c_n N)} \tag{6.26}$$

where N_u = seasonal plant N uptake, $kg\,ha^{-1}$; A_n = maximum seasonal plant N uptake, $kg\,ha^{-1}$; b_n = intercept parameter for plant N uptake; c_n = N response coefficient for plant N uptake, $ha\,kg^{-1}$. Common c would require that $c_n = c$. Now the summing process means using the appropriate A and b values for each data set.

6.2.1 Standard Error

In addition to optimum a_j values for the model, we wish to estimate standard error, δa_j, for each parameter. The procedure now follows. Elements of the Hessian matrix are given by

$$H_{jk} = \frac{\partial^2 E}{\partial a_j \partial a_k} \qquad j, k = 1, 2, \ldots, p \tag{6.27}$$

where p is the number of parameters in the model. For this purpose the Hessian matrix should contain *all* parameters, linear and nonlinear. Variance of the estimate is defined as

$$s^2 = \frac{1}{n - p} \sum_{i=1}^{n} (Y_i - \hat{Y}_i)^2 \tag{6.28}$$

where n is the number of data points. The variance-covariance matrix is given by

$$COV(a_j, a_k) = s^2 [H_{jk}^{-1}] \tag{6.29}$$

Standard error for parameter a_j is then calculated from

$$\delta a_j = \pm (s^2 H_{jj}^{-1})^{1/2} \tag{6.30}$$

where H_{jj}^{-1} is the jth diagonal element of the inverse Hessian matrix. Covariance is given by

$$\delta a_j \delta a_k = s^2 H_{jk}^{-1} \tag{6.31}$$

which gives the cross correlation between a_j and a_k. Ideally the covariance is zero, but it seldom is for real data.

For the logistic model the Hessian matrix is given by

$$[H] = \begin{bmatrix} H_{AA} & H_{Ab} & H_{Ac} \\ H_{bA} & H_{bb} & H_{bc} \\ H_{cA} & H_{cb} & H_{cc} \end{bmatrix} \tag{6.32}$$

where the elements of the matrix are calculated from

$$H_{AA} = \frac{\partial^2 E}{\partial A^2} = 2 \sum \frac{1}{[1 + \exp(b - cN)]^2} \qquad [6.33]$$

$$H_{bb} = \frac{\partial^2 E}{\partial b^2}$$

$$= 2A \left\{ \begin{array}{l} \displaystyle\sum \frac{Y \exp(b - cN)}{[1 + \exp(b - cN)]^2} - 2 \sum \frac{Y \exp 2(b - cN)}{[1 + \exp(b - cN)]^3} \\[4mm] \displaystyle -A \sum \frac{\exp(b - cN)}{[1 + \exp(b - cN)]^3} + 3A \sum \frac{\exp 2(b - cN)}{[1 + \exp(b - cN)]^4} \end{array} \right\} \qquad [6.34]$$

$$H_{cc} = \frac{\partial^2 E}{\partial c^2}$$

$$= 2A \left\{ \begin{array}{l} \displaystyle\sum \frac{YN^2 \exp(b - cN)}{[1 + \exp(b - cN)]^2} - 2 \sum \frac{YN^2 \exp 2(b - cN)}{[1 + \exp(b - cN)]^3} \\[4mm] \displaystyle -A \sum \frac{N^2 \exp(b - cN)}{[1 + \exp(b - cN)]^3} + 3A \sum \frac{N^2 \exp 2(b - cN)}{[1 + \exp(b - cN)]^4} \end{array} \right\} \qquad [6.35]$$

$$H_{Ab} = \frac{\partial^2 E}{\partial A \partial b} = \frac{\partial^2 E}{\partial b \partial A} = H_{bA}$$

$$= 2 \left\{ \sum \frac{Y \exp(b - cN)}{[1 + \exp(b - cN)]^2} - 2A \sum \frac{\exp(b - cN)}{[1 + \exp(b - cN)]^3} \right\} \qquad [6.36]$$

$$H_{Ac} = \frac{\partial^2 E}{\partial A \partial c} = \frac{\partial^2 E}{\partial c \partial A} = H_{cA}$$

$$= -2 \left\{ \sum \frac{YN \exp(b - cN)}{[1 + \exp(b - cN)]^2} - 2A \sum \frac{N \exp(b - cN)}{[1 + \exp(b - cN)]^3} \right\} \qquad [6.37]$$

$$H_{bc} = \frac{\partial^2 E}{\partial b \partial c} = \frac{\partial^2 E}{\partial c \partial b} = H_{cb}$$

$$= -2A \left\{ \begin{array}{l} \displaystyle\sum \frac{YN \exp(b - cN)}{[1 + \exp(b - cN)]^2} - 2 \sum \frac{YN \exp 2(b - cN)}{[1 + \exp(b - cN)]^3} \\[4mm] \displaystyle -A \sum \frac{N \exp(b - cN)}{[1 + \exp(b - cN)]^3} + 3A \sum \frac{N \exp 2(b - cN)}{[1 + \exp(b - cN)]^4} \end{array} \right\} \qquad [6.38]$$

Note again that A, b, and c in these equations are the optimum values from nonlinear regression.

For multiple data sets with individual A and b and common c, the Hessian matrix becomes

$$[H] = \begin{bmatrix} H_{AA} & H_{AA_n} & H_{Ab} & H_{Ab_n} & H_{Ac} \\ H_{A_nA} & H_{A_nA_n} & H_{A_nb} & H_{A_nb_n} & H_{A_nc} \\ H_{bA} & H_{bA_n} & H_{bb} & H_{bb_n} & H_{bc} \\ H_{b_nA} & H_{b_nA_n} & H_{b_nb} & H_{b_nb_n} & H_{b_nc} \\ H_{cA} & H_{cA_n} & H_{cb} & H_{cb_n} & H_{cc} \end{bmatrix} \qquad [6.39]$$

in which the following conditions are assumed

$$H_{AA_n} = \frac{\partial^2 E}{\partial A\, \partial A_n} = \frac{\partial^2 E}{\partial A_n\, \partial A} = H_{A_nA} = 0 \qquad [6.40]$$

$$H_{Ab_n} = \frac{\partial^2 E}{\partial A\, \partial b_n} = \frac{\partial^2 E}{\partial b_n\, \partial A} = H_{b_nA} = 0 \qquad [6.41]$$

$$H_{A_nb} = \frac{\partial^2 E}{\partial A_n\, \partial b} = \frac{\partial^2 E}{\partial b\, \partial A_n} = H_{bA_n} = 0 \qquad [6.42]$$

$$H_{bb_n} = \frac{\partial^2 E}{\partial b\, \partial b_n} = \frac{\partial^2 E}{\partial b_n\, \partial b} = H_{b_nb} = 0 \qquad [6.43]$$

So the Hessian matrix for this case becomes

$$[H] = \begin{bmatrix} H_{AA} & 0 & H_{Ab} & 0 & H_{Ac} \\ 0 & H_{A_nA_n} & 0 & H_{A_nb_n} & H_{A_nc} \\ H_{bA} & 0 & H_{bb} & 0 & H_{bc} \\ 0 & H_{b_nA_n} & 0 & H_{b_nb_n} & H_{b_nc} \\ H_{cA} & H_{cA_n} & H_{cb} & H_{cb_n} & H_{cc} \end{bmatrix} \qquad [6.44]$$

6.2.2   Correlation Coefficient

The nonlinear correlation coefficient R can be calculated from (Cornell and Berger, 1987; Overman et al., 1990c)

$$R = \pm \left[1 - \frac{\sum (Y - \hat{Y})^2}{\sum (Y - \bar{Y})^2} \right]^{1/2} \qquad [6.45]$$

where Y = measured yield, Mg ha^{-1}; $\hat{Y}$ = estimated yield from the model; Mg ha^{-1}; $\bar{Y}$ = mean of the measured yield, Mg ha^{-1}. This subject is explored further in Section 6.7.

6.2.3 Application

The procedure described above for nonlinear parameter estimation is now applied to data from the literature. Creel (1957) conducted an incomplete factorial ($N \times P \times K = 8 \times 3 \times 3$) field study at Gainesville, FL with coastal bermudagrass. Since there was only very modest response to applied P and K, data for each applied N are averaged over P and K for irrigated plots in 1955. Results are given in Table 6.1. Dry matter yields and plant N are normalized to reduce the effect of the scaling factors for each in regression analysis. The second-order Newton-Raphson procedure for individual A, individual b, and common c parameters requires 12 iterations with initial parameter estimates of $b = 2$, $b_n = 3$, and $c = 0.0075$ ha kg^{-1}. Final estimates and standard errors are: $A = 21.4 \pm 0.43$ Mg ha^{-1}, $A_n = 587 \pm 15$ kg ha^{-1}, $b = 1.64 \pm 0.024$, $b_n = 2.32 \pm 0.092$, and $c = 0.0047 \pm 0.00028$ ha kg^{-1}. The overall correlation coefficient is 0.9944.

6.3 PROBABILITY MODEL

The probability model is given by

$$F = \frac{1}{2}\left\{1 + \operatorname{erf}\left[\frac{t - \mu}{\sqrt{2}\sigma}\right]\right\}$$

[6.46]

Table 6.1 Normalized dry matter and plant N response to applied N (1955, averaged over P and K) for coastal bermudagrass at Gainesville, FL

N kg ha^{-1}	Y Mg ha^{-1}	N_u kg ha^{-1}	N_c g kg^{-1}	$Y/21.0$	$N_u/600$
0	2.4	31	12.9	0.114	0.052
62	4.2	48	11.4	0.200	0.080
125	5.8	80	13.8	0.276	0.133
250	9.4	148	15.7	0.448	0.247
500	15.2	326	21.4	0.724	0.543
750	18.1	410	22.7	0.862	0.683
1250	20.4	558	27.4	0.971	0.930
1750	21.7	620	28.6	1.033	1.033

Data adapted from Creel (1957).

where F = normalized yield for the season; t = calendar time since Jan. 1, wk; μ = time to the mean of the distribution, wk; σ = standard deviation of the distribution, wk. The error function is defined by

$$\text{erf } x = \frac{2}{\sqrt{\pi}} \int_0^x \exp(-u^2)\, du \qquad [6.47]$$

Error sum of squares E is defined by

$$E = \sum \{F - \hat{F}\}^2 = \sum \left\{ F - \frac{1}{2}\left[1 + \text{erf}\left(\frac{t - \mu}{\sqrt{2}\sigma} \right) \right] \right\}^2 \qquad [6.48]$$

where F = measured fraction and $\hat{F}$ = estimated fraction from model (Eq. [6.46]). Since both parameters μ and σ occur in nonlinear form, the second-order Newton-Raphson method is used for parameter estimation. Expansion of the first derivatives in the neighborhood of μ and σ leads to

$$\left(\frac{\partial E}{\partial \mu} \right)_{\mu + \Delta\mu, \sigma + \Delta\sigma} = \left(\frac{\partial E}{\partial \mu} \right)_{\mu,\sigma} + \left(\frac{\partial^2 E}{\partial \mu^2} \right)_{\mu,\sigma} \Delta\mu + \left(\frac{\partial^2 E}{\partial \mu\, \partial \sigma} \right)_{\mu,\sigma} \Delta\sigma + \cdots$$

$$\approx \left(\frac{\partial E}{\partial \mu} \right)_{\mu,\sigma} + \left(\frac{\partial^2 E}{\partial \mu^2} \right)_{\mu,\sigma} \Delta\mu + \left(\frac{\partial^2 E}{\partial \mu\, \partial \sigma} \right)_{\mu,\sigma} \Delta\sigma = 0 \qquad [6.49]$$

$$\left(\frac{\partial E}{\partial \sigma} \right)_{\mu + \Delta\mu, \sigma + \Delta\sigma} = \left(\frac{\partial E}{\partial \sigma} \right)_{\mu,\sigma} + \left(\frac{\partial^2 E}{\partial \sigma\, \partial \mu} \right)_{\mu,\sigma} \Delta\mu + \left(\frac{\partial^2 E}{\partial \sigma^2} \right)_{\mu,\sigma} \Delta\sigma + \cdots$$

$$\approx \left(\frac{\partial E}{\partial \sigma} \right)_{\mu,\sigma} + \left(\frac{\partial^2 E}{\partial \sigma\, \partial \mu} \right)_{\mu,\sigma} \Delta\mu + \left(\frac{\partial^2 E}{\partial \sigma^2} \right)_{\mu,\sigma} \Delta\sigma = 0 \qquad [6.50]$$

Equations [6.49] and [6.50] can be reduced to the two simultaneous equations

$$\left(\frac{\partial^2 E}{\partial \mu^2} \right)_{\mu,\sigma} \Delta\mu + \left(\frac{\partial^2 E}{\partial \mu\, \partial \sigma} \right)_{\mu,\sigma} \Delta\sigma = -\left(\frac{\partial E}{\partial \mu} \right)_{\mu,\sigma} \qquad [6.51]$$

$$\left(\frac{\partial^2 E}{\partial \sigma\, \partial \mu} \right)_{\mu,\sigma} \Delta\mu + \left(\frac{\partial^2 E}{\partial \sigma^2} \right)_{\mu,\sigma} \Delta\sigma = -\left(\frac{\partial E}{\partial \sigma} \right)_{\mu,\sigma} \qquad [6.52]$$

This system of equations takes the matrix form

$$\begin{bmatrix} H_{\mu\mu} & H_{\mu\sigma} \\ H_{\sigma\mu} & H_{\sigma\sigma} \end{bmatrix} \begin{bmatrix} \Delta\mu \\ \Delta\sigma \end{bmatrix} = \begin{bmatrix} -J_\mu \\ -J_\sigma \end{bmatrix} \qquad [6.53]$$

where the elements of the Hessian matrix are given by

$$H_{\mu\mu} = \frac{\partial^2 E}{\partial \mu^2}, \quad H_{\mu\sigma} = \frac{\partial^2 E}{\partial \mu \, \partial \sigma} = \frac{\partial^2 E}{\partial \sigma \, \partial \mu} = H_{\sigma\mu}, \quad H_{\sigma\sigma} = \frac{\partial^2 E}{\partial \sigma^2} \qquad [6.54]$$

and the elements of the Jacobian vector are given by

$$J_\mu = \frac{\partial E}{\partial \mu}, \quad J_\sigma = \frac{\partial E}{\partial \sigma} \qquad [6.55]$$

Equation [6.53] can be solved by Cramer's rule to obtain

$$\Delta\mu = \frac{D_\mu}{D} \qquad [6.56]$$

$$\Delta\sigma = \frac{D_\sigma}{D} \qquad [6.57]$$

where the determinants in Eqs. [6.56] and [6.57] are given by

$$D = \begin{vmatrix} H_{\mu\mu} & H_{\mu\sigma} \\ H_{\sigma\mu} & H_{\sigma\sigma} \end{vmatrix} = H_{\mu\mu}H_{\sigma\sigma} - H_{\mu\sigma}^2 \qquad [6.58]$$

$$D_\mu = \begin{vmatrix} -J_\mu & H_{\mu\sigma} \\ -J_\sigma & H_{\sigma\sigma} \end{vmatrix} = -J_\mu H_{\sigma\sigma} + J_\sigma H_{\mu\sigma} \qquad [6.59]$$

$$D_\sigma = \begin{vmatrix} H_{\mu\mu} & -J_\mu \\ H_{\sigma\mu} & -J_\sigma \end{vmatrix} = -J_\sigma H_{\mu\mu} + J_\mu H_{\mu\sigma} \qquad [6.60]$$

Corrections for estimates of μ and σ follow from

$$\mu \text{ (new)} = \mu \text{ (old)} + \Delta\mu \qquad [6.61]$$

$$\sigma \text{ (new)} = \sigma \text{ (old)} + \Delta\sigma \qquad [6.62]$$

Iteration of the procedure continues until

$$\left| \frac{\Delta\mu}{\mu} \right| \le \varepsilon \quad \left| \frac{\Delta\sigma}{\sigma} \right| \le \varepsilon \qquad [6.63]$$

Using the definition

$$x = \frac{t - \mu}{\sqrt{2}\sigma} \qquad [6.64]$$

the necessary derivatives are listed below.

$$\frac{\partial E}{\partial \mu} = -2 \sum (F - \hat{F}) \frac{\partial F}{\partial \mu} \qquad [6.65]$$

$$\frac{\partial E}{\partial \sigma} = -2 \sum (F - \hat{F}) \frac{\partial F}{\partial \sigma} \tag{6.66}$$

$$\frac{\partial^2 E}{\partial \mu^2} = -2 \left[\sum (F - \hat{F}) \frac{\partial^2 F}{\partial \mu^2} - \sum \left(\frac{\partial F}{\partial \mu} \right)^2 \right] \tag{6.67}$$

$$\frac{\partial^2 E}{\partial \sigma^2} = -2 \left[\sum (F - \hat{F}) \frac{\partial^2 F}{\partial \sigma^2} - \sum \left(\frac{\partial F}{\partial \sigma} \right)^2 \right] \tag{6.68}$$

$$\frac{\partial^2 E}{\partial \mu \, \partial \sigma} = -2 \left[\sum (F - \hat{F}) \frac{\partial^2 F}{\partial \mu \, \partial \sigma} - \sum \left(\frac{\partial F}{\partial \mu} \right) \left(\frac{\partial F}{\partial \sigma} \right) \right] = \frac{\partial^2 E}{\partial \sigma \, \partial \mu} \tag{6.69}$$

$$\frac{\partial F}{\partial \mu} = \frac{-1}{\sqrt{2\sigma}} \exp(-x^2) \tag{6.70}$$

$$\frac{\partial F}{\partial \sigma} = \frac{-1}{\sqrt{\pi\sigma}} x \exp(-x^2) \tag{6.71}$$

$$\frac{\partial^2 F}{\partial \mu^2} = \frac{-1}{\sqrt{\pi\sigma^2}} x \exp(-x^2) \tag{6.72}$$

$$\frac{\partial^2 F}{\partial \sigma^2} = \frac{2}{\sqrt{\pi\sigma^2}} x(1 - x^2) \exp(-x^2) \tag{6.73}$$

$$\frac{\partial^2 F}{\partial \mu \, \partial \sigma} = \frac{1}{\sqrt{\pi\sigma^2}} (1 - 2x^2) \exp(-x^2) = \frac{\partial^2 F}{\partial \sigma \, \partial \mu} \tag{6.74}$$

6.4 CONFIDENCE CONTOURS

Procedures for obtaining optimum parameters for the logistic and probability models have been outlined above. These included standard errors of the estimates. Now we take the analysis a step further and examine combinations of the nonlinear parameters which generate equal probability contours. The logistic model is chosen for illustration, but the procedure applies equally well to the probability model. Details have been given by Draper and Smith (1981), Ratkowsky (1983), and Overman et al. (1990b).

The error sum of squares of deviations between the data and the model is defined by

$$E = E(b, c) = \sum \{Y - \hat{Y}\}^2 = \sum \left\{ Y - \frac{A}{1 + \exp(b - cN)} \right\}^2 \qquad [6.75]$$

where $E(b, c)$ is written to emphasize dependence of E on the nonlinear parameters b and c. For minimum error sum of squares, $E = E_0$, for optimum parameter values $b = b_0$ and $c = c_0$, Eq. [6.75] becomes

$$E = E_0(b_0, c_0) = \sum \{Y - \hat{Y}\}^2 = \sum \left\{ Y - \frac{A_0}{1 + \exp(b_0 - c_0 N)} \right\}^2 \qquad [6.76]$$

Now F statistics can be used to develop the approximate relationship (Draper and Smith, 1981) between E_0 and the error sum of squares at some particular probability level

$$E = E_0 \left[1 + \frac{p}{n - p} F(p, n - p, q) \right] \qquad [6.77]$$

where p = number of nonlinear parameters; n = number of observations; q = probability level; and $F(p, n - p, q)$ = F statistic.

6.4.1 Application

Overman et al. (1990b) applied this procedure to data from Beaty et al. (1964) for bahiagrass grown at Americus, GA. Average yield data with standard deviations for the period 1959 through 1963 are listed in Table 6.2 and shown in Fig. 6.1. Yield estimates are from the logistic equation

$$Y = \frac{10.77}{1 + \exp(1.05 - 0.0123N)} \qquad [6.78]$$

Table 6.2 Average yield response to applied N of pensacola bahiagrass grown at Americus, GA (1959-1963).

Applied N kg ha^{-1}	Mean yield, Mg ha^{-1}	
	Measured	Estimated
0	2.77 ± 0.32	2.79
45	3.81 ± 0.30	4.08
90	5.91 ± 0.50	5.54
180	7.98 ± 0.70	8.21
360	10.47 ± 0.83	10.41

Yield data adapted from Beaty et al. (1964).

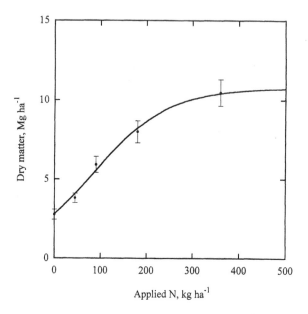

Figure 6.1 Dry matter response to applied N for pensacola bahiagrass. Averages and standard deviations for five years (1959–1963). Data from Beaty et al. (1964) as discussed by Overman et al. (1990b). Curve drawn from Eq. [6.78].

where the parameters were obtained by nonlinear regression. The corresponding minimum error sum of squares is $E_0 = 0.2624$. Now consider the linear parameter fixed at the optimum value $A_0 = 10.77$ Mg ha^{-1} and focus on the dependence of E on b and c. Then for $p = 2$ with $n = 5$, at the 95% probability level we have F(2, 3, 95) = 9.55. Substitution of these values into Eq. [6.77] leads to

$$E = (0.2624)[1 + \tfrac{2}{3}(9.55)] = 1.93 \tag{6.79}$$

With these results, Eq. [6.76] becomes

$$\sum \left\{ Y - \frac{10.77}{1 + \exp(b - cN)} \right\}^2 = 1.93 \tag{6.80}$$

The challenge now is to find combinations of b and c which satisfy Eq. [6.80]. The approach is to fix a value for parameter b and use a search method to obtain parameter c which satisfies Eq. [6.80]. Combinations of b and c which satisfy Eq. [6.80] can then be used to draw an equal probability contour as shown in Fig. 6.2. The most probable estimates and

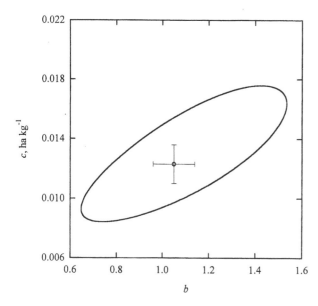

Figure 6.2 Confidence contour for logistic model parameters b and c at 95% confidence level for pensacola bahiagrass from Overman et al. (1990b). Also shown are best estimates and standard errors of b and c.

standard errors are $b = 1.05 \pm 0.090$ and $c = 0.0123 \pm 0.0013 \, \text{ha} \, \text{kg}^{-1}$, which are also indicated in Fig. 6.2.

6.5 SENSITIVITY ANALYSIS

In model analysis it is frequently desirable to determine sensitivity of the model to changes in parameter values. For example, in the example discussed in the previous section for the logistic model the most probable values of the parameters are $A = 10.77 \, \text{Mg} \, \text{ha}^{-1}$, $b = 1.05$, $c = 0.0123$ $\text{ha} \, \text{kg}^{-1}$. The corresponding standard errors of the estimates are $\delta A = \pm 0.37 \, \text{Mg} \, \text{ha}^{-1}$, $\delta b = \pm 0.09$, and $\delta c = \pm 0.0013 \, \text{ha} \, \text{kg}^{-1}$. One approach is to change one parameter by the standard error while holding the other parameters at optimum values and calculate changes in the output. Overman et al. (1990b) discussed results for this example. Change in A causes a change in the plateau of the response function. Change in b causes a change in the intercept at $N = 0$, while the effect for parameter c produces mostly change at intermediate ranges of N. Another approach is to calculate change in the error sum of squares E, resulting from change in a single parameter, typically around the optimum value. The change could be by the standard error of the estimate. One could also use Eq. [6.77] to estimate E for a particular probability level. For small changes near the optimum parameter value, a plot of E vs. parameter will approximate a parabola.

There are other alternatives for evaluating performance of a model. A scatter diagram between estimated and measured output is frequently used, as shown in Fig. 6.3 for the example above, where the regression line is given by

$$Y(\text{estimated}) = 0.090 + 0.988\,Y(\text{measured}) \qquad\qquad [6.81]$$

In this case the ideal would be a perfect 45 degree line with no scatter; the smaller the scatter the better the correlation. A variation of this is a residuals plot of difference between estimated and measured output vs. estimated output. In this case one examines the plot for trends as well as the amount of scatter. Overman et al. (1991) have discussed both of these approaches for the logistic model.

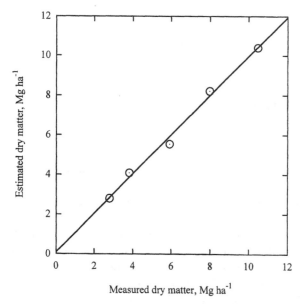

Figure 6.3 Scatter diagram of estimated vs. measured dry matter for pensacola bahiagrass from Overman et al. (1990b). Line drawn from Eq. [6.81]

6.6 DIMENSIONLESS PLOTS

A technique frequently used in science and engineering is that of dimensionless plots. This can serve either of two purposes. One is to reduce a large amount of results into a compact form, and the other is to test the general applicability of a model to a physical system. Let us focus on the latter of these. If a mathematical model happens to describe a given set of data well, it could be claimed that this is an artifact for that particular set of data. The question is how to compare different sets of data where parameter values may differ in order to evaluate the general structure and form of the model. Overman et al. (1990a) utilized this technique to compare results from four field studies with bermudagrass for the logistic model and showed high levels of correlation among variables. This should add greater confidence to the model in question.

Let it be said that students often find dimensionless graphs and equations distasteful when first encountered. It looks like a way for authors and teachers to unnecessarily complicate things and make them less intelligible. My advice is to become accustomed to this approach, because the advantages generally outweigh the disadvantages. In the modeling of physical phenomena a number of dimensionless groups (such as Reynolds, Mach, Damköhler, etc.) have been identified as key to characterizing various processes. Overman et al. (1988) applied this approach to process analysis of overland treatment of wastewater.

6.7 CORRELATION COEFFICIENT

The correlation coefficient R is often used as a measure of "goodness of fit" between a model and data. It lies in the range $-1 < R < +1$. An alternative measure is the coefficient of determination, which is the square of the correlation coefficient. Cornell and Berger (1987) give an excellent discussion of this subject for both linear and nonlinear models.

Many statements have been made about R values. A student once told me that his R value was good, since it was greater than 0.5. On the other extreme, a reviewer once wrote that R was a useless measure for nonlinear models, even when the value exceeded 0.99. In another case, I had to defend and document my use of four digits to report $R = 0.9985$. This was viewed as an overkill, and should only be reported to two decimal places, in which case my value would have been 1.00. However, there was enough scatter apparent in the graph for a reader to see that the correlation was not "perfect." This poses a dilemma. My argument, in which I prevailed with the associate editor, was that we are comparing R to 1, not 0. Therefore, a

value of 0.9985 represents two significant digits different from 1, and that
should be the criteria.

My conclusion is that the correlation coefficient can be a useful measure
of the quality of fit of a model to data, but should be viewed along with
other indicators, such as standard error of the estimate of parameters.

EXERCISES

6.1 For the data of Beaty et al. for bahiagrass given in the table below
    a. Plot yield vs. N for each year on linear graph paper.
    b. Calculate standardized yield for each N and year from

$$Y^* = Y[1 + \exp(1.05 - 0.0123N)]$$

    c. Calculate average and standard deviation from (b) for each year.
    d. Discuss some possible causes of the variation from year to year in
       (c).
    e. Write the equation for the 75% probability confidence contour for
       parameters b and c for this system.
    f. Discuss results of this analysis.
    g. What level of uncertainty in estimating forage yield and nutrient
       uptake do you feel is acceptable in practice?

Dry matter yield of pensacola bahiagrass
at Americus, GA

Applied N kg ha^{-1}	Seasonal dry matter, Mg ha^{-1}				
	1959	1960	1961	1962	1963
0	2.55	2.78	3.21	2.92	2.39
45	3.87	3.62	3.47	3.82	4.26
90	5.96	5.63	6.71	5.83	5.40
180	7.44	7.32	8.03	8.03	9.09
360	10.24	9.84	11.93	10.14	10.20

Data adapted from Beaty et al. (1964).

6.2 Creel (1957) measured dry matter accumulation with time for coastal
bermudagrass on Leon fine sand (sandy, siliceous, thermic Aeric
Haplaquod) at Gainesville, FL. One set of data is given in the table
below, where time is referenced to Jan. 1. Maximum dry matter for the
season is estimated as 6.20 Mg ha^{-1}.
    a. Plot F_y vs. t on probability paper.

b. Use the graph to estimate the mean and standard deviation from $\mu = t\ (F = 50\%)$ and $\sigma = [t(F = 84\%)\ t(F = 16\%)]/2$.

c. Perform nonlinear regression on F_y vs. t to obtain optimum estimates of μ and σ. Use values from (b) as initial estimates in the second-order Newton-Raphson procedure.

d. Draw the best fit line on (a) with the optimum parameters.

e. Plot Y vs. t on linear graph paper.

f. Draw the best fit curve on (e) from the model.

g. Discuss the results of the model on the data.

Dry matter accumulation with time (irrigated, $N = 125$ kg ha^{-1}, averaged over applied P and K) for coastal bermudagrass at Gainesville, FL

t wk	ΔY Mg ha^{-1}	Y Mg ha^{-1}	F_y
	1.00		
15.0		1.00	0.161
	1.02		
21.0		2.02	0.326
	1.23		
26.9		3.25	0.524
	0.73		
30.9		3.98	0.642
	1.75		
37.4		5.73	0.924
	0.36		
47.7		6.09	0.982
—		6.20	1.000

Data adapted from Creel (1957).

REFERENCES

Adby, P. R., and M. A. H. Dempster. 1974. *Introduction to Optimization Methods.* John Wiley & Sons, New York.

Bard, Y. 1974. *Nonlinear Parameter Estimation.* Academic Press, New York.

Bates, D. M., and D. G. Watts. 1988. *Nonlinear Regression Analysis & Its Applications.* John Wiley & Sons, New York.

Beaty, E. R., R. G. Clements, and J. D. Powell. 1964. Effects of fertilizing pensacola bahiagrass with nitrogen. *J. Soil & Water Conser.* 19:194–195.

Cornell, J. A., and R. D. Berger. 1987. Factors that influence the value of the coefficient of determination in simple linear and nonlinear regression models. *Phytopathology* 77:63–70.

Creel, J. M. Jr. 1957. The effect of continuous high nitrogen fertilization on coastal bermudagrass. PhD dissertation. University of Florida, Gainesville, FL.

Draper, N. R., and H. Smith. 1981. *Applied Regression Analysis.* John Wiley & Sons, New York.

Freund, R. J., and R. C. Littell. 1991. *SAS System for Regression.* 2nd ed. SAS Institute, Cary, NC.

Hosmer, D. W., and S. Lemeshow. 1989. *Applied Logistic Regression.* John Wiley & Sons, New York.

Kleinbaum, D. G. 1994. *Logistic Regression: A Self-Learning Text.* Springer-Verlag, New York.

Overman, A. R., E. A. Angley, T. Schanze, and D. W. Wolfe. 1988. Process analysis of overland treatment of wastewater. *Trans. Amer. Soc. Agric. Engr.* 31:1375–1382.

Overman, A. R., E. A. Angley, and S. R. Wilkinson. 1990a. Evaluation of a phenomenological model of coastal bermudagrass production. *Trans. Amer. Soc. Agric. Engr.* 33:443–450.

Overman, A. R., A. Dagan, F. G. Martin, and S. R. Wilkinson. 1991. A nitrogen-phosphorus-potassium model for forage yield of bermudagrass. *Agron. J.* 83:254–258.

Overman, A. R., F. G. Martin, and S. R. Wilkinson. 1990b. A logistic equation for yield response of forage grass to nitrogen. *Commun. Soil Sci. Plant Anal.* 21:595–609.

Overman, A. R., C. R. Neff, S. R. Wilkinson, and F. G. Martin. 1990c. Water, harvest interval, and applied nitrogen effects on forage yield of bermudagrass and bahiagrass. *Agron. J.* 82:1001–1016.

Ratkowsky, D. A. 1983. *Nonlinear Regression Modeling: A Unified Practical Approach.* Marcel Dekker, New York.

Seber, G. A. F., and C. J. Wild. 1989. *Nonlinear Regression.* John Wiley & Sons, New York.

Index

9 780367 395896